A Short Sunderland flying boat moored on the Thames. In order to let it through, the bascules of Tower Bridge had to be opened (Crown Copyright).

Aviation Enthusiasts' Guide to
LONDON &
THE SOUTH-EAST

Aviation Enthusiasts' Guide to
LONDON &
THE SOUTH-EAST

Peter G. Cooksley

PSL Patrick Stephens, Cambridge

First published 1982

British Library Cataloguing in Publication Data

Cooksley, Peter
    Aviation enthusiasts' guide to London and the
    South-east.—(Action stations)
    1. Aeronautical museums—England
    I. Title II. Series
    629.133'025'4222   TL506.G7

    ISBN 0 85059 533 9

Text photoset in 10 on 11 pt English Times
by Manuset Limited, Baldock, Herts.
Printed in Great Britain on 100 gsm coated
cartridge, and bound, by The Garden City Press,
Letchworth, Herts, for the publishers,
Patrick Stephens Limited, Bar Hill, Cambridge,
CB3 8EL, England.

Contents

Introduction

The number of men and women who are interested in aviation, whether it is through business, sport or relaxation, seems to increase in size at a steady rate and, to meet the needs of their enthusiasm, there are many excellent books which are intended for serious study; and even the author must plead guilty to having increased the number of these works. However, there are far fewer books that may be classed in the same group but with the label 'for light reading'—a situation which seems wrong, for the aviation enthusiast is as entitled as any one else to enjoy his lighter moments. Additionally there exists, all over the country, a huge quantity of historical reminders of our aviation heritage and this surrounds us, often undetected, in our everyday lives. These two factors are here brought together in a book which cuts across the accepted works on many aspects of the aeronautical world and it is natural that the capital should be chosen for this first book on the subject.

The field from which to draw has been a vast one and it would have presented few problems to prepare a book of double the length. Inevitably, therefore, the choice of subject must be the author's. In making this, experience gathered in meeting aviation enthusiasts visiting the area covered, from abroad and other parts of the country, has been drawn upon plus the fact that a mixture of well-known and lesser-cherished associations have been included. With regard to the former it has been remembered that man's conquest of the air has gathered, like everything else, its share of legend, often resulting in a contradiction of facts, and so care has been taken to set down only the authentic or, when the mists of time have proved too thick, the version of a particular legend which seems the most likely. As an extra precaution against perpetuating the doubtful, those events which have taken place within living memory have only been accepted from sources which might have claim to some degree of local or residential knowledge.

Naturally, in order to achieve an even spread of interesting facts in a book of this size—bearing in mind that a volume could be compiled for every county in the south-east—it seems that some appealing and even obvious pieces of local legend have had, reluctantly, to be left out. Therefore no claim is made to please

Left *Part of the Battle of Britain Memorial Flight, this Lancaster bomber is seen in the markings of No 44 Squadron over the Victorian Embankment in connection with the 1976 Royal Tournament* (Crown Copyright).

every reader 100 per cent. However, it is confidently believed that the balance is in large measure redressed by the inclusion of fresh material.

It is obvious that a work of this nature could never have been compiled without the aid of a large number of helpful and enthusiastic people, agencies and references and I am indebted to many for factual and photographic assistance including the following: Professor Harold M. Barlow; Keith Belcher; Malcolm Butler, Press Officer, STC Limited; Mrs Mary Cobley; Peter D. Cornwell; Norman W. Cruwys; The Editor, *The Croydon Advertiser*; Leslie J. Dickson; Brian Downham; The Promotions Manager, Fortnum & Mason; Victor J. Garwood; Alfred E. Jessop; Peter Lamb; Miss C.H.N. Leigh; Edwin T. Maidment, JP; Mrs Doris McDonald; F.A.W. Mann; H.E.J. Monk; Rev J.H. Read; R. Rignall; Bruce Robertson; The Secretary, the Royal Air Force Club; Patrick W. Saunders, MA; R.A.R. Wilson, Public Relations Manager, British Airways.

Prologue

It was King Bladud who started it all but, on that fine morning of 843 BC, it is doubtful if his legendary Majesty realised that his pioneer attempt was to be the first of a long series of connections with the art of aerial navigation which the capital was to know. At the moment he was far too occupied with other matters, including the simple one of keeping his balance, for the morning was somewhat breezy and the playful little gusts that whirled about the 'high eminence' which had been chosen as a take-off pad, not only sent cold shivers up the short royal tunic but also ruffled the feathers of the great bird that were strapped to his back.

Below, the crowd which had gathered were also pioneers in their way for they were the first citizens to converge on a certain spot to see an air show, although it must be admitted that some of them had also come in the hope of seeing something gruesome with which they could later entertain their friends. Not that Bladud was unpopular, far from it, he had managed to keep his throne for 20 years and from his palace at Trinavantum, later to be called London, nine generations of his family, not counting himself, had ruled Britain. Perhaps one reason why his subjects tolerated him with a knowing nod of the head and a smile when his name was mentioned was that he was apt to burn the midnight oil following a strange hobby. He was keenly interested in necromancy with the result that visitors waiting to be ushered into the royal presence sometimes found it a little difficult to keep their breakfasts down as odd sounds and smells wafted through the corridors. Give him his due, though, they all agreed, he had done more than a bit of good, for had he not discovered the curative powers of Bath water when drunk after meals? However, this must have caused some unexpected results in an age when few could write, and information had to be passed on by word of mouth, a means of communication which still does not indicate the use of capital letters, all-important in this case. Others merely announced that any king was a great monarch who advocated being drunk after meals and yelled for another bottle of mead.

However, with the wind now becoming playful to the point of rudeness, Bladud had more immediate problems for, only a short distance off, loomed the bulk of the Temple of Apollo and from the unexpected viewpoint it looked very large and very solid. Yet in the circumstances there was little that could now be done about it for, not only were the people below becoming restless and pointedly going off from time to time in little groups to consult the sun dial, but the royal prince was also hovering in the background, anxious to see his dad off the parapet. His Majesty had always tried to be a good example to his heir although it had been difficult at times for young Lear could occasionally be a bit strange and was far too democratic with the Court Jester as well. But all that was his mother's fault—here was a chance to show the lad how a full-grown man handled a situation like this.

Nevertheless there was still the problem of the Temple in the way and beyond it the river, but this at least promised a cushioned landing for much of the surrounding land was soft, if smelly, marsh. Anyway had not the spirits, which the king had conjured up to foretell the future, assured him that all would be well and that they would lend a hand—be with him in spirit as it were? He was not skilled in the necromantic arts for nothing. Whether it was because he had finally made up his mind or it was to rid himself from the odd flea that had emigrated to warmer climes from the disused eagle's wings, will never be known, but, at this point, the king was seen to square his shoulders and then turn into wind. It was the work of a moment to deposit a little of the royal saliva on the royal finger and raise the august digit to determine wind strength and direction, then with a turn of the head and a farewell wink at young Lear, who by now seemed more interested in the view of a well-appointed serving maid which his elevated position afforded, the King jumped.

The tale of this first London flight was set down in 1148 by Geoffrey of Monmouth, that celebrated writer who was the author of *Chroniconsive Historia Britonum*. His ecclesiastical training had made Geoffrey a kindly man so that we must accept with reserve his description of what followed, for the wild flapping that took place is dignified by the term 'flight'. Whatever your attitude to this there can be no question that Bladud must have consulted the spirits about his future on one of the royal 'off days', for they certainly let him down. Indeed they did so far too sharply, right on the Temple of Apollo in fact, breaking the royal neck and thus making the king the triple first of Britain; first flyer, first air crash and first fatality.

The saga spun by the Benedictine chronicler of the flying king (who would have lived at much the same time as Elijah) was no more than a legend. However, tradition and fact united in a strange way early this century when fragments pre-dating the first Christian church were found beneath the Roman brick in the foundations of Westminster Abbey, traditionally supposed to have replaced Apollo's Temple after an earth tremor in 150 AD in the time of Antoninus Pius. Be that as it may, Bladud is unique to London even though he belongs to such flights of fancy as Pegasus who was adopted as a device by the Inner Temple in 1563 or Robert Palstock's flying women of 1751 who were supposed to have navigated an early form of hang glider. From the kit that they were wearing it might seem that summers in bygone days really were better than in our own age since this consisted of no more than the lower part of an abbreviated bikini and a hairband. The resultant nudity no doubt ensured increased airflow efficiency. All this shows that London's aviation heritage may therefore be easily found . . . .

Chapter 1

Around Victoria

One of the entrances to central London is the railway station in Belgravia and it is only necessary to approach the front to find (to the left) the first of our aviation connections. This is the large and decorative archway which gives on to what was, at one time, termed the 'continental' side. Today it looks probably better than it has done for many years as a result of the cleaning and renovation which was carried out at the end of the 1970s. It is the fact that this side of Victoria had always been associated with the boat-trains and departures for France that created the connections with the armed services which this entrance has. Although, nowadays, few of the hurrying passengers (many of them students or holiday-makers, who swarm in both directions through the portals, and are bowed down with the weight of luggage and the cares of travel) spare a thought for the significance of this archway.

The British Caledonian office at Victoria Station maintains the terminus' long connection with flying.

This photograph of the main entrance to Victoria Station was only able to be taken after the buildings in the way were demolished.

Between the years 1914 to 1918 armed men marched through the archway in thousands. The roof has thrown back a million times the echo of the studded boots as men of the Royal Flying Corps and other soldiers (for until the last year of the First World War they were a corps of the Army) in addition, of course, to the Royal Navy and the RNAS, passed through here to entrain for the British south coast ports. For many of them this arch was the last they ever saw of London—for their bones still lie in France and other foreign fields, and seas. It is therefore with truth that some have termed this entrance 'the gateway to Valhalla'.

Another connection which the area has with aviation, and this time a more direct one, is to be found in the history of the same period of time. To be exact it was the closing weeks of 1916 and, as the hands of the clock crept towards noon on the morning of Tuesday, November 28, the people of London were going about their everyday business unmindful that they were about to become the focal point of a piece of history. Certainly they had become accustomed to the fact that war had assumed a new dimension and that their homes now offered none of the security which had been enjoyed by the island race since the last of the raiding Danes had been repulsed in the time of King Alfred. Human beings, though, owe their survival as a species to their adaptability. Raids from the air had now become part of a way of life and there was a certain feeling of optimism, for the airships which came to drop their loads had proved more vulnerable than had at first been believed. Only a little over 12 weeks earlier the first of these night raiders to be downed over Great Britain had fallen at Cuffley so that something of the terror had gone out of the echoing boom of the warning maroons fired from the police stations. However, the 'All Clear' sounded by Boy Scout buglers (as the two notes 'G-C'), as they careered through the streets on their bicycles, was none the less welcome, accompanied as it often was by a 'cock-a-doodle-do' sounded on railway locomotive whistles.

Now it was a hazy morning in early winter, but one that was to pass into history as the first daylight attack by an aeroplane on London. This daring sortie was to be carried out by two German flyers, Deck-Offizier Paul Brandt, pilot, and Leutnant Walther Ilges, the observer. They were flying an LVG CII described, for some reason at the time, as 'an LVG Type D 9'. This was a frail biplane incapable of a maximum speed in excess of 81 mph although the ceiling of about 13,000 feet was ideal for an operation of this nature.

With this sort of altitude making it inconspicuous above the broken winter cloud, the first of the 10 kg bombs dropped. It hurtled down as a complete surprise at exactly 11.50 am to explode a little east of Brompton Road on what the press was to describe, in slightly patronising tones, as 'a humble dwelling house', causing injury to one of the persons within. As the LVG continued its course towards the railway station, the remainder of the stick of bombs was released, amounting to six in all. One of these fell on the rear of a house roof which it penetrated before exploding so that some damage was done inside. Contemporary reports speak of the hole in the slates looking 'as if a giant hand had punched its way through from the inside'.

Meanwhile people in the streets, unless they were in the immediate vicinity of the explosions, were largely ignorant of what was going on and those who did hear the noise of the bombs attributed it to other causes. 'I thought there had been a gas explosion', afterwards observed one woman who had been in a shop at the time. The only injury which was claimed to be of any severity was suffered by a victim of the third bomb of the salvo—a woman in a house about 100 yards further to the east.

Further hurt was caused to a woman cleaner (although this was slight) who was in another house when a bomb struck the roof, blowing in the windows in the surrounding area and smashing the coping stones. But it was the final bomb, dropped just before the brave and audacious flyers circled Victoria Station itself, which caused the most comment, perhaps because the effect was the most spectacular. This fell in a cobbled roadway so that the walls of surrounding buildings were severely chipped by flying splinters, a water pipe was burst and steel shutters were, for good measure, 'peppered' by the blast.

The whole attack had lasted exactly 10 minutes and, despite the slight time spent over the railway junction, pilot Paul Brandt now had to think about the problem of getting home because his machine was limited to a mere four hours' flying and quite obviously fuel and precious minutes may have to be squandered if it should be necessary to fight their way out. In this he was very near the truth for, even at that moment, no less than 17 British interceptors were climbing into the November sky.

The course to London where, it was later alleged, the Admiralty building was the prime target, had, for the first part, followed the line of the Thames Estuary and it was not until Sheerness had appeared under the port wings that a turn had been made over land. This had taken the LVG north of Strood where another turn had been made, this time due south, skirting Wrotham and then round in a gentle curve to the west which was to take the flyers south of Redhill in Surrey.

Once again the course had been changed, this time north, leaving Croydon on the starboard side.

With the twin objects of the sortie now completed, namely the dropping of some bombs and the exposure of several plates in the aerial camera that was carried, it was a case of making as much progress in as short a time as possible and, steering almost exactly south-east now, the little two-seater passed north of Bromley and continued until Wrotham was reached again. Once more this was skirted on its southern side but thereafter the Deck-Offizier and the Leutnant gave every sign of having temporarily lost their way, for they were next fleetingly seen proceeding east in the direction of Maidstone.

As with other built-up areas, this was carefully skirted and it seems that here the navigational error was realised for the pair passed almost right round the town before making off in the direction of Tunbridge Wells. Avoiding this on their starboard beam, the airmen then set course for the English Channel so that they passed over the coast between Hastings and Bexhill-on-Sea, passing within only a short distance of the latter. Their luck had held well for at no time had any of the defending aircraft been rewarded by so much as a sight of the intruders.

The newspapers of the following day had evidently decided to play down the whole affair and any readers who had been on the receiving end of the attack in the localised areas, where the presence of the enemy aeroplane had been made clear, were to be disappointed. Perhaps it was the spirit of the times, perhaps before the days when electronic communication had not yet rendered the daily printed word a hopeless anachronism, that editors did not feel the need to resort at every occasion to the largest type in a form of desperate rear-guard action. The fact is, the news media on the 119th day of the war's third year decided, in modern parlance to 'play it cool' and, on an inner page, *The Times* contented itself with the following brief report: 'A Midday Raid on London. Yesterday morning an enemy aeroplane, flying high above the haze, dropped six bombs on London. Nine persons (sic) were injured; one, a woman, seriously. The material damage was slight'.

It will probably never be known just how serious the destruction was and the report is not entirely accurate in reporting hurt to only nine persons for a total of ten is now generally agreed, but the fact remains that much later an announcement was made that damage to the value of £1,585 was caused—a considerable sum in that day, so that one must suspect that more than a few tiles and windows suffered.

As is well known, the aero engines of that age were notoriously fickle and, up to now, the 160 hp Mercedes motor which had served Paul Brandt and Walther Ilges so well began to show signs of malfunction, just at the precise time when it was most needed—for, below, the French anti-aircraft batteries were throwing up some fierce and fairly accurate fire. The flyers held their height for as long as possible but plainly their long-held luck had at last run out and eventually there was no alternative than to admit defeat and carry out a forced landing in enemy territory. It was not long before excited Frenchmen were all around, mounting guard with fixed bayonets, examining the machine as if it had come from another planet and delving into the cockpits. Here they found the 20 plates which had been exposed over London, showing, it was claimed, bomb damage and a map; some said of the British capital's anti-aircraft defences.

Understandably, the French gunners imagined that the LVG had fallen a

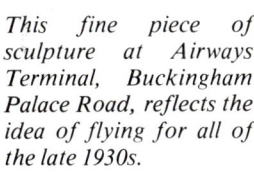

This fine piece of sculpture at Airways Terminal, Buckingham Palace Road, reflects the idea of flying for all of the late 1930s.

victim to their fire and claimed it as such when the machine had finally succumbed at 2.15 pm, so that a newspaper was able to comment in its Editorial the next day: 'There appears no reasonable doubt that the raider has paid the penalty, the honour here falling to our French Allies'. However, it was the final comment that showed how prophetic the writer was when he added: 'this isolated visit is by no means to be ignored'. How right he was, the public consternation was reflected in a letter received from Mr Francis Bennet Goldney, writing from the House of Commons. This supplies us also with a nice reflection on the efficiency of the British postal service of the day as, even though they were operating under the trials of war-time conditions, the letter arrived at the offices of *The Times* so that it was able to be printed in the first edition of the issue for the day following the airmens' intrusion!

The aeronautical connections which may be found about Victoria Station are by no means exclusively connected with war, as a trip up Buckingham Palace Road, west of the station, will show. Little more than a quarter of a mile along here, just beyond Elizabeth Street, stands the massive block, dominated by its tower which can be seen by train passengers entering the station, known as Airways Terminal. As the style of the large piece of sculpture over the main entrance indicates, the present building dates from the late 1930s but, at the time, it was rather smaller. The two wings at either side of the central block were opened in August 1965 as part of a £3 million scheme of modernisation and enlargement. This enabled some of the departments of British Overseas Airways, formerly scattered around the West End, to be housed in the central building, although some of them had to be accommodated opposite Semley Place over the new eight-storey block erected by London Coastal Coaches.

The central block of Airways Terminal was finished at the end of 1938 and was opened the following year. It was designed to replace former offices used by Imperial Airways on the continental departure side of Victoria Station which had served from June 1931, having previously occupied a site in Charles Street alone. A feature of the new building had been the now-vanished decor of the entrance foyer which was carried out exclusively in 'Empire' hardwoods and these were even applied to the massive fluted columns of the main hall where the veneers had been carefully chosen for the quality and beauty of their contrast. A final touch to give the impression of luxurious space, as was expected of air travel in that day, was supplied by the blue rubber floor. With the main counters situated immediately opposite the entrance, one end of the concourse had been taken up by a serpentine-shaped bar while the restaurant had been housed in the basement.

The reason for the choice of site is interesting since the idea hinged on the fact that long distance air travel to countries in the British Empire was carried out largely by means of the flying boat. The advances in load-carrying capacity (which military aviation demanded of designers) made possible the high-capacity landplane, although at the penalty of the huge aerodromes we know to-day.

The idea, therefore, of siting the new Imperial Airways building alongside the line of the (then) Southern Railway was that passengers for the distant parts of the Commonwealth (a name just coming into use) could check in at the Buckingham Palace Road Terminal and then catch the train from the special platform behind the main building. In one swift and efficient movement they could be whisked from the reception point to that of departure for the last word in luxury air travel of the day, namely the famous 'Empire' Class flying boats, plying from Southampton with night-stops *en route.*

In order to fit the building for the new demands of the jet age, the changes we see today were carried out, as has been stated, although the handsome blocks of white Portland stone specified originally for the frontage were left untouched. This was, in fact, the third major change which had been carried out to the building since 1939. It was only the following year that the newly-formed BOAC took over the building, as from April 1, at the same time as it acquired the former Imperial and British Airways' Corporations. It was to better accommodate the passengers of the same organisation, now operating in the post-war years, that the Airways Terminal received an interim modernisation in 1958.

The changes we see today were, however, much more far-reaching and included the removal of all the fine timber of the old scheme, largely because, attractive as they were, they did nothing to lighten the appearance of the concourse. This was entirely replaced by facings of white 'crystal' marble on both columns and walls, thus greatly increasing the atmosphere of light and space. This was added to by the renewal of the flooring with a mosaic in two shades of grey and blue and the lighting was improved by the introduction of the rows of closely-spaced fittings we now see.

Other changes consisted of the removal of the extra seating which had been introduced in 1958 to cope with the increased numbers of passengers who had to be handled in the post-war world. This seating was shifted to the mezzanine floor which had itself been extended into the new wings and had new office accommodation installed round its periphery. On the lower floor the old basement restaurant still exists but is now used by the staff. Instead there is a new dining room and a similar one reserved for private functions. The doors and partitions are now of plate glass and this is also used for the balcony balustrade so that it provides a contrast to the panels of green marble which are inserted at intervals on the columns of the foyer.

The Dornier 17Z which crashed into Victoria Station during the Battle of Britain caused this damage to the stonework that may still be seen.

All these and other adaptions to fit Airways Terminal for the demands of modern air travel were carried out by the contracting firm of Sir Robert McAlpine & Sons working to the designs of Sir Thomas Bennett & Son. The latter were the consultant architects, working in conjunction with the former BOAC properties branch, who wisely decided to leave the exterior of the building to present very much its original appearance, except for the substitution of logos on the tower in the space beneath the clock originally occupied by the Speedbird of the pre-war concept. The proposed sale of this building was announced in January 1981.

Back in Terminus Place, in front of Victoria Station and near to the junction with Wilton Road, it calls for only a casual examination of the stonework to discover that this part of the station wall has at some time sustained damage from some form of impact and the reason for this once more provided another of the capital's links with the aviation world. It was Sunday, September 15 1940, and the air fighting that has since become known as the Battle of Britain was in full swing in the skies of Southern England. In France the Luftwaffe bomber crews were once again preparing for the hated crossing of the Channel. Among the more experienced of them to climb into the morning mist, with its promise of clearance later, were the young men in Dornier 17Z, *2361*, with the markings on its side—F1 FH—which declared it to be part of 1/KG76 based at Nivelles, just a short distance south of Beauvais. The time was 10.05 am.

At the controls was the 27-year-old pilot, Oberleutnant Robert Zehbe. Near him, crouched over the radio, sat Obergefreiter Ludwig Armbruster with Leo Hammermeister; the remaining members of the crew being Gustav Hubel and Hans Goschenhofer, the observer. All three were Unteroffiziers, and the crew was well experienced having taken part in some other of the attacks which had been ordered a little earlier against British fighter stations. This Sunday morning the target was more inviting, nothing less than London itself. Now, formating with the remaining members of the Staffel over Cap Griz-Nez, it was possible to see from 15,000 feet that the anticipated clearance in the haze was already taking place.

While the formation was heading for landfall at Dungeness the energetic Sergeant Ray Holmes of the RAF, until only a short time before a newspaper reporter on the staff of his father's Liverpool agency, was enjoying a quick bath and hardly had he got in than one of his colleagues began beating on the door, yelling that 504 Squadron was being placed at readiness. If the Sergeant had read the injunctions in the newspapers of that summer it is unlikely that he was very wet for war-time economies constantly urged one to make do with only four inches of water. Be that as it may, with a bound, Ray Holmes was making an attempt to dry himself and fling on some clothes at the same time, before rushing out to the waiting vehicle which was to career off, bumping and lurching across the Hendon airfield while he pulled his socks on as the motor went along.

On the northern side, the car stopped with a jerk and even as the pilots tumbled out, the Tannoy was blaring an order for the Hurricanes to scramble, the final words of the announcement being all but drowned as one after the other the Merlin engines roared into life, sending great clouds of steel-blue smoke billowing back across the airfield as the ground crews fired the motors into life. It was only moments later that the 12 Hurricanes, with the throttles fully open, were climbing to the rendezvous height of 12,000 feet over Hendon, while below the few cars that were out on that Sunday morning looked like toys as they made their seemingly snail-like progress down the Edgware Road.

The fighters of No 504 were not the only ones to have scrambled, however, to intercept the Dorniers. A little earlier other Hurricanes had also been streaking across the grass of their airfield at Duxford near Cambridge, the base of the Czechoslovakian Squadron, which, despite the distance that they had come from the area of No 12 Group, now had the advantage of an earlier alert and had the bombers of KG76's First Staffel in sight over south London. Meanwhile all was not going entirely well for Robert Zehbe and his crew for one motor refused to give full power so that they were slowly being left behind by their swifter colleagues. Now, added to their mounting problems, there was anti-aircraft fire to contend with although, in the event, they suffered no damage.

The main body of the formation was still in sight when the first Hurricane came in to the attack at a few minutes after mid-day, despite the continued fire from the London defences below. Hardly had Flight Lieutenant Jefferies, leading 'B' Flight, taken his thumb from the gun button and banked aside to allow the remainder of his section to bring their sights to bear, than flames were seen to be licking from under the Dornier's port engine. But still the bomber refused to go down, despite the attentions of Sergeants Puda, Hubacek and Kaucky. Shortly afterwards they were joined by two more Hurricanes, this time from 609 Squadron and it was at this moment that a pair of parachutes were seen to blossom below the Nazi machine indicating that, while they still had the advantage of 3,000 feet of altitude, the radio operator and mechanic regarded the bomber as doomed.

A short time later, Robert Zehbe had been able to shake off his attackers and was managing to keep the main formation in sight when out of the empty air

Oberleutnant Robert Zehbe's Dornier, minus outer wing panels and rear fuselage, plunges to its doom on Sunday, September 15 1940 (Imperial War Museum).

appeared the fighter with Sergeant Holmes at the controls. First he fired at the port wingman and when he saw that both of its engines had stopped, he transferred to the other side and fired a burst at the winged Dornier there. Almost at once it was his turn to be surprised by the appearance of a parachute as the German pilot followed the example of his two colleagues.

Meanwhile, Ludwig Armbruster was landing in Sydenham, where he was arrested in Wells Park Road only half a mile from the former site of Crystal Palace and mechanic, Leo Hammermeister, who was wounded, came down on the west side of Dulwich. Robert Zehbe was less lucky. He landed at Kennington and was immediately attacked by a howling mob of civilians who took advantage of his wounds to harm him grievously before he could be rescued by the police so that he later died in hospital.

Believing that it was a gunner who had baled out, since the Dornier seemed still committed on its course, Holmes made up his mind to ram the bomber since his ammunition was now low so, jockeying for position, he brought his Hurricane round in a great scything arc so that his wing severed the tail from the bomber where the fuselage was slimmest. But the Sergeant's elation at what he had done was short lived for it quickly became clear that the dive into which the manoeuvre had thrown the Hurricane was becoming steeper and it refused to pull out. The dive became steeper until, almost vertical, it turned into a spin. There was only one thing for it, Ray Holmes must follow the example of his victims and it was the work of a moment to thrust back the hood of *P2725* and allow the force of the spin to toss him from the cockpit although there had been a nasty moment after he had unfastened his harness when the parachute pack fouled the cockpit equipment.

The Dornier was now going down in another part of the sky, or what was left of it, for the rear fuselage, forward of the point from which the tail had been hacked, suddenly broke off, containing the dead body of gunner Gustav Hubel, and fell away over Fulham. A moment later the wings went, too, leaving the main part of the fuselage with the wing centre section and its engines to scream earthwards over Victoria Station.

Sergeant Holmes was also over the same railway terminus, hanging from his parachute straps minus boots which had been lost in the exit from his fighter. His concern now was that while Hyde Park, from where a number of people had seen the event, spread inviting and open to the north he knew that the lines below him, where he must land, were electric and to these he swung ever closer. The ground seemed to rush up to him now and one of the right-hand swings seemed to direct him beyond the lines and, in a moment, he realised that he was to be saved from electrocution only to be dashed on the roof of a block of flats across Ebury Bridge Road near the junction with Buckingham Palace Road.

The force with which he struck the tiles knocked the breath from Holmes for a moment but self-preservation asserted itself and, suddenly, he was sliding down the rake of the roof, clawing for a hold. In a flailing bundle of arms and legs he went over the gutter and shot past the windows. Then, just as the paving seemed to rush finally towards him, he stopped with a jerk that dug the parachute harness painfully into his flesh. He looked down and was surprised to find that he was all-but standing in a dustbin, his feet prevented from quite reaching the bottom because he was now suspended a few inches from the ground by the silk canopy which had snagged a drainpipe. There was no one about to see him work the harness release strapped across his stomach and step gingerly from the refuse in his socks.

Despite the seeming quiet of the Sunday afternoon, up the road in front of Victoria Station where Terminus Place joins Wilton Road, there was plenty to attract the crowds for the bulk of the Dornier 17Z had finally come to earth against the shop of James Walker, the jewellers, while the broken tail unit had smashed down nearby on a roof in adjoining Vauxhall Bridge Road. It was the first enemy aircraft to crash on Central London.

Chapter 2

Wings over the city

The name of Moorfields to our generation means the excellent eye hospital. But in the second half of the 18th century it meant something completely different for, at that time, the area was sufficiently open for the district to offer space for the exercise fields of the Honourable Artillery Company—that ancient body which, even in 1783, had enjoyed the privileges of the Charter granted to them by Henry VIII for 246 years.

The architect of the sensational events which were to be presented to the public on this spot was a sailor of fortune who had studied science and mathematics at Parma before enlisting in the Spanish Navy, where he saw action off the coast of America and against the Moors. The future seemed bright for the young Count Francesco Zambeccari until he fell foul of the hated Spanish Inquisition and he fled to political asylum in England, eventually ending up in Cheapside where he had a friend.

It was from here that he launched the first balloon which Londoners had seen when he decided to emulate the Montgolfier's experiments on November 4. Despite the small diameter of the envelope, less than two metres, its voyage to Waltham Abbey, Essex, where it landed 13 miles from its starting point, was marked by excited crowds gazing skyward in the streets of London, the largest gathering taking place in Highgate, where the famous hill offered a particularly good vantage point. In the simpler world of that day, showmen were quicker to take advantage of public acclaim so that, exactly three weeks later, the new spectacle brought Zambeccari to the HAC ground with a balloon double the diameter of the first.

The launch of this one was to be public and great crowds assembled to see the unmanned envelope cut free, and a short time later many were all agog to learn that landfall had been made 48 miles distant after a voyage of two and a half hours at Graffham, near Petworth, Sussex. Until now, few had thought of the idea of a man acting as passenger in such a balloon and the Count once more gained public attention when he announced that such an experiment was to be carried out from Hyde Park in February of the following year. Now his luck seemed to desert him, for financial troubles forced the project to be abandoned and Francesco Zambeccari at once did as he had before and boarded a ship to take him to find fresh conquests.

There was, however, the handsome 26-year-old from the Neapolitan embassy, Vincent (Vincenzo) Lunardi, who had similar ideas and, in his search for a suitable lift-off site, wrote to the Governor of Chelsea Hospital, in the first

The Lyceum Ballroom, still retaining something of its theatre-like appearance, marks the approximate spot where Lunardi's balloon was constructed.

instance, for permission to use the grounds there. This was granted, subject to 50 per cent of the profits being donated to be shared between a pensioner's family and insurance against any damage to the buildings.

The balloon for this venture was being built 'at the Lyceum'—the precise location described as 'near Exeter Change', Exeter Street, off the Strand which is still in existence today. To see the vessel under construction, both single and season tickets were issued to the public who could view the work at any time between 10 am and 8 pm. The idea of this arrangement was to finance the actual flight but, alas, the expected support was poor and, to add to Lunardi's troubles, permission to use the Chelsea site was withdrawn at the last moment due to the activities of disappointed hooligans after an abortive attempt to launch a balloon by another aeronaut on an earlier occasion. This was not the greatest of Lunardi's problems, for, with takings nothing near the expected £1,200, the builders threatened to impound the balloon.

The new venue chosen was the obvious one of the HAC grounds at Moorgate and it was not until the day before the date announced for the demonstration that the new hydrogen balloon was finally released from the Lyceum and, in order to supervise the inflation, Lunardi sat up all night. September 15 1784 dawned fine and clear so that it was not long before the first of the 20,000 spectators began to arrive at the rural spot on the outskirts of the city. A little later the throng was joined by the Prince Regent for whom a silken canopy had been erected over a special stand and he was joined by Burke, Fox and Pitt from among the politicians of the day.

By noon the greater part of the 18,200 cubic feet of gas had been pumped into the envelope and now a new problem arose, for it was evident that, despite the diameter of about 10 metres, the balloon was incapable of lifting Lunardi together with his friend, George Biggin. An hour later the envelope was fully inflated and George Biggin, despite his financial support for the venture, had agreed that, since his own skill as a balloonist was very limited, the honour for the flight must go to the Neapolitan and so, the latter, having shaken hands with the Prince, boarded the car with a pigeon, cat and dog for company and the lines were cast off. An atmosphere of awe and even fear was evident among the crowd from the complete silence which prevailed; no one waved or raised a cheer and the general feeling was manifest when the Prince of Wales was seen to raise his hat as the wind carried the vessel towards the east, the wind later veering and taking a more northerly direction.

It was now that Lunardi discovered that the pair of oars he had brought along were useless as a means of propulsion horizontally, although he seemed to still believe that they in some way governed the vertical movement of his vessel. Their use did, however, help to keep him warm for although the altitude at this point in the adventure was about 1,000 feet, far below the four miles be believed himself to be at, the temperature was minus 16 degrees Centigrade, he claimed.

There were those who whispered that the whole affair of balloon voyages was little more than a personality stunt by the brave Neapolitan but there is no doubt that he had some first rate qualities. Among these was his kindness and an example of this is his concern for the cat which was becoming increasingly distressed by the cold. At about 3.30 pm the red, yellow and blue-stripped balloon was over a cornfield at North Mimms, some 13 miles from the point of departure and Lunardi decided to do something about his feline passenger. He therefore 'rowed' down to where a group of 25 people were standing and gave the cat into the care of an elderly lady who fearlessly approached the visitor from the clouds. One of the dozen men in the group was later to prepare a long sworn statement describing what he had seen that Wednesday afternoon, adding that he could not approach too closely as his horse was restive but going on to say that after the balloon had risen again he had heard the gentleman in the car calling out farewell through a 'trumpet'.

In order to rise, Vincent Lunardi had thrown out the last of his sand and, to gain an even higher altitude, knives, forks, spoons and food were likewise abandoned so that the balloon, now adorned with a necklace of frozen moisture below the envelope, rose to such a height that the voyager was moved to write down a description of the beautiful panorama below during the next half hour's flight.

It was now that the second descent was made, this time to a cautious reception at Standon, near Ware in Hertford, and it fell to a young girl to shame the male spectators when she came forward alone and set an example by grasping one of the lines and soon other willing hands were holding the balloon down on to the grass of Long Mead where a monument was later set up recording the event. The rough-hewn boulder still stands with an inscription which sums up the attitude of wonder that was to mark public opinion of early balloon flights.

'Let posterity know and knowing be astonished, that on the 15th day of September 1784 Vincent Lunardi of Lucca in Tuscany, the First Aerial Traveller in Britain, mounted from the Artillery Ground in London and traversing the Regions of the Air for two Hours and fifteen Minutes, in this spot revisited the

Earth. On this rude monument, for ages be recorded that wondrous Enterprise successfully achieved by the Powers of Chemistry and fortitude of Man, that Improvement in Science which the Author of All Knowledge, patronyzing by His Providence the Invention of Mankind, hath graciously permitted to their Benefit and His own Eternal Glory.'

Presented at Court, created an honorary member of the Artillery and with a silver medal struck in his honour, Lunardi announced the use of a larger balloon for the ascent to be made on May 13 1785. In place of the stripes which Nathaniel Whitbread had so carefully noted for his statement, the envelope now bore the Union Flag and with George Biggin and the beautiful Mrs Letitia Sage, a popular actress, he hoped to rise once more from the Artillery ground. Alas, events were to repeat themselves for so many could not be lifted and Lunardi had once more to voyage alone.

Once aloft, all seemed to be going well when, for some unexplained reason, probably a gas leak, the balloon rapidly came down without warning and landed heavily in Tottenham Court Road where the unfortunate occupant discovered that the British flag he waved to reassure those on the ground as he descended was no safeguard against rough handling. He never flew from the city again but, instead, used a site on the other side of the river (see Chapter 3).

Perhaps at the backs of the minds of those who treated Lunardi so badly there lurked a subconscious thought that the way was being opened for invaders from the clouds which would forever destroy their security. Be that as it may, the fact remains that the First World War was to leave many a mark on the ancient square mile.

It was natural that the river should act as a magnet for all sorts of activity and one structure connected with the watery highway, Tower Bridge, came in for a great deal of activity which still continues on occasion. Two years, almost to the day, before the outbreak of the cataclysm in Europe, found a young man, Frank McClean, fascinated by the frame provided by the upper and lower parts of the bridge. It was between these, the upper of which is only 142 feet above the level of the water, that he flew his Short biplane on August 10 1912. The remaining bridges demanded different techniques but the problem was partly resolved by the fact that the machine was a floatplane. The remainder of the journey up river to Westminster was accomplished either by hopping over the obstructions or by taxiing under them and the aircraft was finally brought to rest in the shadow of Big Ben. It was not long before the conduct of this young man, who was among the earliest holders of a Royal Aero Club pilot's certificate, brought a police reprimand in reply to which he promised not to fly from the water again until he had taxied down river. This he did, taking Tower Bridge in a more conventional manner this time but, having cleared it, his attempt to take off ended ignominiously in a crash.

Lieutenant-Colonel Sir Francis McClean, as the youthful flyer later became, would have completely understood the next assault on the Wolfe-Barry/Jones structure, one that was to stir up few strong feelings for, by now, the 'war to end wars' was over and, as 1919 was being written on the datelines, high spirits and daring had temporarily come back into fashion. The performer this time was a Flight Lieutenant S. Pickles of the RNAS and the machine to be used was what the contemporary press termed a Fairey (patent) variable wing seaplane—the Fairey Hamble Baby (of which by this time fewer than 18 remained in service) although they had been used in some numbers.

Take-off had been made from Isleworth at 10.40 am and the flight along the Thames began at a speed of 120 mph, it was claimed, seemingly somewhat ambitious for a machine powered by a 130 hp Clerget rotary motor, at an altitude of no more than 1,000 feet, and Pickles lost height as Tower Bridge came into view. To a pilot skilled in the control of his machine the complete span does not present a very formidable challenge at first for there is little sensation of 'going through the eye of a needle' since, with the fixed portions of the roadway, the complete span is some half a mile across. Even so, with the smallness of comparative headroom, it is not a feat for those who lack confidence and with these facts in mind the Lieutenant carefully lined up his small machine. As the distinctive rhythm of the rotary motor echoed for a moment from the narrow confines of the river banks with an odour of burnt castor oil in the nostrils of the watchers, the little 28 foot hornet was through and climbing steadily away.

Interviewed later, Lieutenant Pickles was to claim that machines of the type which had been used were specially constructed to counter the threat of the German Brandenburg seaplanes. Although it is correct to confirm that 'several more were available when the Armistice was signed', history does not seem to corroborate the envisaged special use nor the subsequent statement that there was a possibility of seaplanes being employed for a London postal delivery. This service was only flirted with some 30 years later with a pair of red-painted helicopters which pioneered the world's first helicopter night mail service in East Anglia on February 21 1949 with Sikorsky S-51s. Meanwhile it was to be several decades before the next and final attempt was made on Tower Bridge and other bridges, this time by Kit Draper, 'the Mad Major'.

However, this is not to say that there are no more tales to tell of the famous span and aeronautica, for there remains one more, that is largely forgotten today. It springs from the day when the designers of parachutes were giving attention to the problem of escape from aeroplanes at a low altitude, although in this they were some distance ahead of the official attitude in this country which frowned on the whole idea of pilots being equipped with a means of escape, the whole thing being summed up in the declaration which is on record that if parachutes were issued pilots might be tempted to leave their machines if the going became too hot!

It was a civilian, the retired engineer, E.R. Calthrop who brought a more intelligent note to the whole problem by inventing the well-known and rather unfortunately-named Guardian Angel parachute. This, although far from perfect, was superior to the Spencer type which was issued to the observers of balloons operating at the Front. The Calthrop type was designed about two boards between which the canopy was packed, rather in the manner of the filling in a sandwich roll. On being released, the silk then opened over the lower board.

By November 1917 it had been decided to test a couple of Guardian Angels from a low height since extended drops had been carried out and had proved successful, at the surprisingly early date of May 1914. This was when Vickers had co-operated in trials at Barrow-in-Furness the previous year, having seen a drop from an airship at 7,000 feet altitude. Rather strangely the venue for the new low-level trials was situated in the heart of war-time London, the exact spot

Left *One of the demonstration leaps from Tower Bridge with the aid of a Guardian Angel parachute in November 1917.*

Right *In more recent years the bascules of the bridge opened to admit a Short Sunderland flying boat* (Crown Copyright).

being Tower Bridge which afforded a height of little more than 140 feet above the Thames.

Since the Guardian Angel was not a free-fall parachute, but one which worked on the static line principle, for the purposes of the demonstration the upper walk of the bridge was fitted with a platform and protruding arm for the suspension of the sandwich-like pack. From here two men jumped and, despite the fact that the drop of less than 200 feet did not permit the canopy to open fully, safe landings were made in the Thames on both occasions.

Alas, there seems to have been little impression made on the ignorance of certain executive factions even by the happy outcome of these demonstration trials. This was a particularly dismal result since the Parachute Committee was meeting at about the same time. It was a body where only the Secretary had any unprejudiced knowledge of the subject to be examined so that, when the Commander of the 3rd Brigade, RFC, supported the idea of parachutes, with the suggestion that the method of operating a Calthrop could easily be met by fitting the packs on the turtle-backs of aircraft, the whole matter was simply side-stepped.

However, with the advantage of hindsight one is tempted to wonder if the chilly dip in the Thames which the two suffered was completely wasted effort. In January of the following year, the Air Board suddenly placed an order with the Calthrop manufacturers and lifted the ban on advertisement, an embargo which even the contemporary press seemed determined to uphold, despite the highly public place of the experiments, for reports of them are singularly conspicuous by their absence in the more important publications of the day.

For completeness one must cast the aeronautical net over not only the city

proper but also over the immediately surrounding areas for there are no barriers in the sky and thus it is that Farringdon Road comes next into our tale. The tall business premises that look down on the famous street-vendors barrows, dealing exclusively in books, give the appearance of long endurance without change but, as a large plaque at pavement level records, one, number 61, is strictly-speaking not as old as its fellows.

On the night of September 8 1915 two airships crossed the coast, *L13* and *L14*. The latter came in a few miles west of Cromer and made off in the direction of Dereham which it circled before setting course for the coast again and passing directly over Sheringham before turning for home. Its companion crossed the coast between Wells and Hunstanton and then set course more or less due south, passing west of King's Lynn and directly over Cambridge before making for London. This line of approach had been completely different from those adopted by three airships on the previous night when *SL2, LZ74* and *LZ77* had all made landfall over Essex, bombing Chelmsford among other targets.

Despite what the newspapers and other vendors of propaganda had to say it is extremely unlikely that the raiders had much idea as to their exact whereabouts so that it was more by accident than intent that bombs were dropped on and around the City of London, *L13* releasing a total of 15 high-explosive and 55 incendiary bombs. Of the former, one severely damaged a premises in Addle Street while another blasted buildings in nearby Bartholomew Close, at the same time bursting a granite fountain which thereafter was much photographed as a curiosity.

It was evidently a bomb from the same stick that, as the first light of the early September morning was just beginning to tint the sky in the east, destroyed the building in Farringdon Road. When it was re-erected a short time afterwards, having been restored in every detail to its former appearance, despite the demands of the worsening war situation at that time, the builders were able to leave for posterity something of the bewilderment and horror felt by the generation which was to be the last brought up in the centuries-old belief in the security of an island people, with the words that may still be read; 'These premises were totally destroyed by a Zeppelin raid during the World War on September 8th 1915. Rebuilt 1917.' By the time that the builders had finished their work there the situation had changed as much as had the attitude of the people, for aeroplanes had replaced the vulnerable dirigibles and were now putting in an appearance in daylight.

Between these dates, strides had been made in the establishment of a proper London defence system and, although planned in the summer of 1915, largely to appease a public outcry (Servicemen back from the Front described the results of the early raids as 'negligible'), there is no doubt that the Zeppelin raids of September 7/8 did much to speed the formative work.

In consequence it was as early as only two months later that, probably with an eye to calming public fears, the Royal Naval Anti-Aircraft Mobile Brigade paraded in the rain their five weapons plus a searchlight during the Lord Mayor's Show on November 9. In subsequent years this same procession was to provide a valuable means of maintaining the morale of the city when captured

Far left *The airship* L13 *dropped bombs on the night of September 8 1915 which destroyed the building on this site (61 Farringdon Road).* **Above** *The plaque which recorded the destruction.*

Left *This is the only building in Bartholomew Close which survived the air raids by airships in the First World War and the night 'blitz' of the Second World War.*

Below *Another survivor of the attacks on the City during the First World War, when several fires were started on July 7 1917, is this one near Telegraph Street.*

booty was displayed and when guns and an LVG CII were towed through the streets, the latter calling forth from a reporter the rather smug comment that '. . . its mechanism was as a sealed book to the majority', 1916 was described as 'singularly rich'.

Although its effect was not exclusively confined to the City area, which had now suffered damage on two occasions to Liverpool Street Station, the audacious raid in daylight by 22 Gothas on July 7 1917 is of particular interest since we have a recollection of it. It was a Saturday morning and a very small girl living in the vicinity of Hendon Aerodrome thought nothing of the sound of aero-engines as she lay in bed late in order to allow her mother to get some cleaning done downstairs. But it was for only a few more moments that young Doris' daydream was to last for she was suddenly aware of her mother bursting into the bedroom and whipping her little daughter from between the sheets in what seemed a single movement. 'It's a raid!', she explained, taking the child to the window so that a somewhat bleary-eyed gaze could attempt to focus on the 22 specks that were even at that moment passing close to the aerodrome. They had seemingly dropped their loads and were now making for home since their course was more or less due east at the time.

Little Doris wanted to watch the excitement particularly as there was no frightening gunfire, no opposition at all in fact, but her mother insisted that they hide under the stairs, while all the time she feared that a knock would soon come at the door and bring the dreaded news that her husband, following the pattern of the age and working in the City for the Saturday morning, had been

Left *Twenty-two Gothas dropped bombs on a Saturday morning in 1917. Some fell near Liverpool Street Station when this bus, B804, standing outside, was damaged* (Imperial War Museum).

Right *The pillars of Liverpool Street Station still bear scars from the occasion on September 10 1940 when the terminus received a direct hit with a 100 kg bomb.*

killed. However, the little child could not accept this and was later to go out and buy her father some handkerchiefs as a 'welcome home' gift, financed from her own savings.

When he did return, quite unscathed after lunch, he had a tale to tell of the events he had seen. Several of the offices adjacent to his had been destroyed and one remarkable incident had been his sight of what seems to have been a 50 kg bomb striking the surface of the road and skidding along horizontally before finally exploding a short way off, killing a horse and destroying the cart to which it was harnessed. This was the same raid during which the redoubtable James McCudden found himself flying alongside the raiders in the hope of his Sopwith Pup distracting the attention of enemy gunners so that his colleagues could close in to an attack. This role of 'groundbait' turned out to be somewhat hazardous when a well-aimed burst of fire from a Gotha carried away his windscreen so that the next few moments found him attempting to increase the distance at which he flew, powerless, since all his ammunition was exhausted, so the strange formation continued in the skies over Southend.

By now Liverpool Street Station had been hit by bombs on one occasion and this was to be later repeated. It was also to suffer again in the Second World War when, on the day when raiding had been temporarily reduced to the nuisance level, September 10 1940, a 100 kg bomb scored a direct hit on the terminus. In fact, the damage to the main structure of the station was limited, but even today there remains for those who wish to seek them out, the marks of bomb splinters on the pillars of Platform 2. Unfortunately for the historian,

Above *This Naval airship* (R26), *seen here over the City under misty conditions, had previously flown over Trafalgar Square in connection with the sale of War Bonds on November 9 1918* (Imperial War Museum).

Right *The waste and inefficiency of airship passenger flights was highlighted by the speed and comfort offered by aeroplane travel as typified by this Imperial Airways advertisement of 1931.*

time has erased all evidence, seemingly, of the earlier ordeal when hits were scored by three bombs on the station itself.

The peaceful years brought a few aeronautical connections which have left their mark to the city at first. One that left its impression only on men's minds, reviving memories of the earlier years when the simple airships used for coastal patrol provided photographers on several occasions with a whole set of field days when their bulks flew over St Paul's, was the excitement created on October 4 1930 by the appearance of the R101 airship. London had seen her before during the same year when she had appeared in connection with the Air Pageant at Hendon, but now she was on her maiden scheduled flight, the first stop being scheduled in Egypt.

Her release for use had been reluctant since there had never been a full speed test and she had not been tried out in heavy weather. Now it was a little after 8 pm but her lights made it easy for those few who stood in the streets to see her, despite the rain. Some said she was already wallowing and some were merely impressed. Warning of deteriorating weather had certainly been received on board by radio and the captain held his course over the Channel while the guest passengers enjoyed dinner in the luxurious lounge before adjouring for cigars in the smoking room—only a matter of feet from the five million cubic feet of

·· Air travel is best!

British supremacy in airplane construction and performance is demonstrated indisputably by the great liners of Imperial Airways —the British Air Line. Wherever you want to travel—to any part of the Continent, to India or to Africa—go by Imperial Airways. It will save you time and trouble.

Save time and travel in comfort
by IMPERIAL
AIRWAYS
THE BRITISH AIR LINE

Full particulars and reservations from any Travel Agency, or
direct from —
IMPERIAL AIRWAYS LIMITED, Airway Terminus, Victoria Station
(Continental Departures) LONDON, S.W.1.

Telephone:
Victoria 8242.
(Night and Day)

Telegrams:
Impairlim,
London.

C.F.H. 101

hydrogen which gave R101 her buoyancy. With the night, when all except the crew had gone to their staterooms, the storm worsened, the sky was black, the sheets of rain icey and the winds made progress extremely difficult.

Over France a few passengers were wakened by the new note of the straining motors but even at that moment the vessel was refusing to answer to the helmsman. There was insufficient time to do much about the situation now, ballast was seen to drop but it was only for a moment that the bulk of the envelope righted itself. Suddenly a great light filled the sky, a rumbling explosion followed and all that was left of the great vessel which had excited Londoners and those in the south of England only hours before, burnt out on a hillside at Beauvais. It was left to a silver-haired French poacher whose attention had first been attracted to the airship by the lights going out twice before the end to pronounce the valediction with Gaelic directness. 'It went up in the air', he said, 'and there was an end to it. Fini! fini!'

A total of 46 perished immediately in the crash and their remains were brought back to London to lie in Westminster Hall on October 10 while an unending stream of Londoners filed past from 8 am until midnight. RAF men with black armbands stood with arms reversed and bowed heads beside the coffins; and a small boy remembers the deep impression left on his mind when the radio news bulletin, to which his mother was listening, was begun by the announcer stating that, in the manner of what was then called Armistice Day, there would be a two minutes silence before the news was continued.

By now the city had its own RAF Squadron, appropriate for an area which had been the spot selected for the exhibition (in October 1916) of parts salvaged from the early German airships destroyed in the war. The Headquarters of the HAC at Moorgate was chosen, and one of the exhibits had been the observation car of LZ90. It is still being examined by Londoners, as it is now in the Imperial War Museum at Lambeth. Nine years later No 600 Squadron of the Auxiliary Air Force was formed, later with a badge incorporating the sword from the City of London arms. The establishment of this unit took place at Northolt on October 14. Its first role, one that was retained for nine years, was that of a light-bomber squadron. This was changed in July 1934 when it had converted from its ex-war-time DH9as, and later Hawker Hart bombers, to Demon two-seat fighters.

By this time the base had been, since 1927, the aerodrome at Hendon. One change which had not taken place had been any alteration in the source of its personnel which continued to be drawn largely from the city banks, its offices and the Stock Exchange. It seemed entirely appropriate, therefore, that the first machines operated should bear the white shield of the city on their sides, a custom continued on the fins of the next type to be taken on charge, the Westland Wapitis in 1929. The squadron later became unique in having the right to two badges, that already mentioned and the official one granted by King George VI in 1944, when the motto *Praeter sescentos*—'More than six hundred', that is, 'A Great Number', was adopted.

The first mobilisation of the city's own squadron took place at the time of the Munich crisis in September 1938, and Kenley in Surrey became its new base. With Neville Chamberlain's 'peace in our time' declaration, No 600 was stood down and, in the time gained, began to convert to Bristol Blenheim fighters, once more at Hendon. Then, in September 1939, these were flown to Northolt and this time there was to be no stand-down and the City of London Squadron was embarked on a war that was to take it via Redhill and Manston, among others, to such distant bases as Luqa, Souk el Khemis and Campoformido, where it was disbanded in August 1945 only to be reformed in July of the following year at Biggin Hill. Her Majesty Queen Elizabeth became its Honorary Air Commodore until the final disbandment nine years later.

Apart from those already described there are few immediately recognisable traces of aviation's associations with London's city to be found today, if one discounts the unhappy days of the war years from which it is still possible to see here and there the manhole cover bearing the white distinguishing letter of the service it covered. These signs of a past age in the pavements are diminishing now. In the skies there have been many such reminders, not only for the square mile but for London as a whole, with the once traditional fly-pasts that took place annually on the anniversary of the Battle of Britain's 'greatest day'. The first of these was carried out on September 15 1945 when it was led by Douglas Bader in company with, said the press of the period, other 'top scorers'. This was really the first of a new ritual as previously fly-pasts had been only occasional and usually connected with royal events.

Five years later, when the sight of the 'lone Hurricane and Spitfire' in the lead had become part of the accepted custom, an Air Ministry Order was issued giving authority for similar displays on 'Occasions of national importance' by a minimum of three squadrons, so that the King's official birthday of that year was marked in this manner during June when 18 Lancaster bombers performed the duty. Naturally, the Coronation of our present Queen was, like her Jubilee, another occasion when the mounting of a fly-past was regarded as appropriate but the veteran participation in the 'greatest day' celebrations were to come to an untimely end, but not over central London, so that what happened will be related in its proper place elsewhere.

One of a diminishing number of Second World War marks on manhole covers which give access to underground services. This one bears a white 'E' for electricity.

Above *Bristol Sycamore* WT924 *lands on the South Bank near to Hungerford Bridge* (Crown Copyright).

Below *A preserved Gloster Meteor,* WH364, *exhibited outside RAF Kemble, Gloucestershire, in the colours of No 601 City of London Squadron.*

In 1948, the City was directly involved in a new sphere of aviation for, from a National Car Parks site, laid waste by bombing (in the shadow of St Paul's Cathedral), a Bristol 171 helicopter, *VL958*, set the ancient walls vibrating as it clattered into the air at the beginning of the 'Hare and Tortoise' race. This was organised by *The Aeroplane* magazine which was instrumental in bringing the capitals of England and France within 47 minutes of each other.

Thus the associations went on and continue to do so. A few years ago the Fleet Air Arm Museum at Yeovilton in Somerset mounted an exhibition in the Royal Exchange and, in continuance of the theme, on November 6 1977, the few who took a stroll in the vicinity of the Cathedral on that Sunday morning were surprised to see a policeman holding up the traffic in order to permit a Naval Seafire, *SX137*, again from Yeovilton, to be wheeled into place across Cannon Street in Old Change Court to mark the association with Naval aviation of the new Lord Mayor, Sir Peter Vanneck, who had earlier in his career flown this self same type over a period of four years.

Those who would seek some tangible sign of the first year that the City was assaulted by bombs—the attacks only began in May 1915—should retrace their steps and look at the front gate to St Bartholomew the Great. This had been tile hung for so many countless ages that men had forgotten how it looked, perhaps at the time when the youthful Richard II faced Wat Tyler's mob in its shadow, until a bomb from a Zeppelin tore away the ancient disguise to reveal the medieaval timber and plaster beneath.

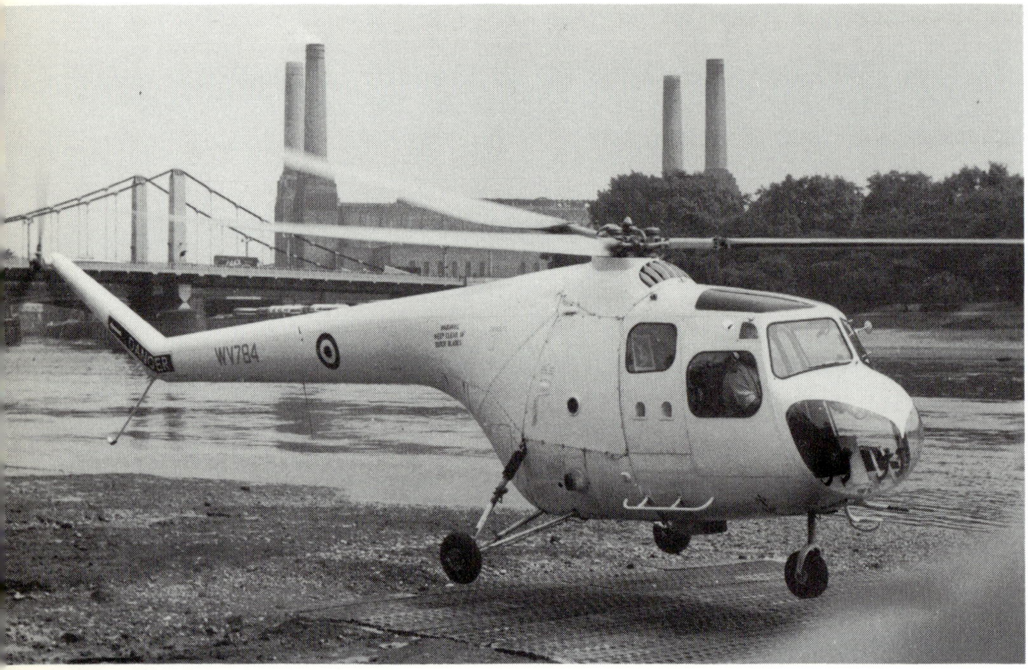
WV784
DANGER

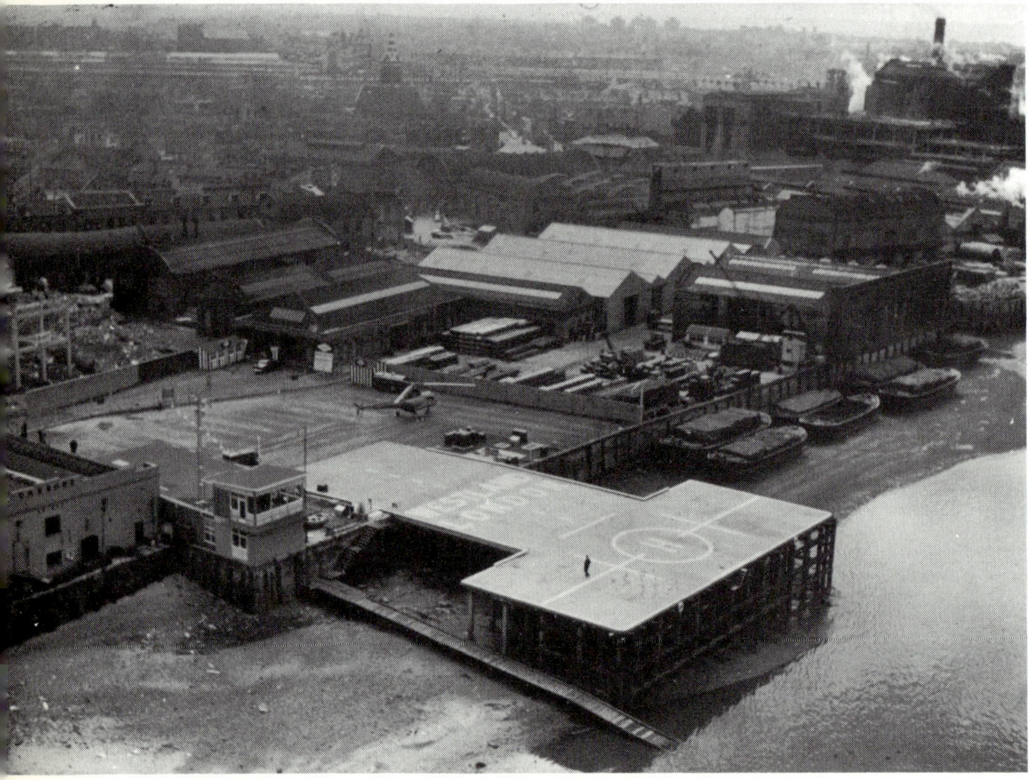

Chapter 3

Along the Thames

London's river, a highway since ancient times, has attracted, not only its own share of aeronautical connections but also the important centres which have accumulated on its shore, these too have added to the richness of the collection. As we have already seen, Tower Bridge acted at various times almost like a magnet to airmen but, on September 30 1931, it was not this span alone which was to be the focus of attention. Major Christopher Draper, DSC, C de G, on this occasion, not only piloted his de Havilland Puss Moth through the gap at the Pool, but also flew through the central arch of Westminster Bridge; a somewhat hair-raising feat as may be imagined today by anyone who cares to stand on the pavement *above* and imagine they are *looking down* on a diminutive aeroplane!

By now officialdom had caught up with things and the use of the Thames as a landing and take-off area was regulated by Air Navigation Directions which demanded that permission be granted for this sort of thing and flights below an altitude of 1,000 feet. No such leave had been granted to Kit Draper, with the result that he was summoned forthwith.

Twenty-two years later the redoubtable Major experienced the same urge, despite his 61 years. On May 5 1953 he set off again to repeat, and even improve on, the earlier escapade. His intention was to fly under all the bridges from the Tower to Kew, and there are 24 of them! This time the machine was an Auster which was carefully brought down almost to the level of the water and, in this manner, 15 of the bridges were taken. Alas for 'the Mad Major', despite the frequent declarations that Coronation Year was the herald of a new age of Elizabethan adventure, the Air Navigation Directions were not seen to embrace an atmosphere of panache. Draper, who had undoubtedly brought his machine below 1,000 feet, was once more summoned and fined.

However, this is not to say that the Thames was not to be used by aerial craft in any way, for the present day finds, in the shadow of the Battersea Power Station, the Heliport with the same name. The route to this is that of the river's course, so that the noise of the rotors brings the minimum inconvenience to Londoners, coming, as it does, from machines dutifully keeping above the minimum altitude. One wonders how many people even realised that the landing

Above left WV784 *lands on the side of the Thames with Battersea Power Station as a background* (Crown Copyright).
Left *'Westland—London' reads the legend on the landing pad of the London Heliport.*

The bulk of the Power Station once again forms a background to the Heliport, this time with a Westland Widgeon, G-ANLW, *making a touch-down.*

pad was there before some television reporters, a short time ago, brought it some unexpected publicity by the discovery that one of the pilots who regularly used the facilities beside the river happened to be a girl, Anne McMunnies. No doubt the beginning of all this can be traced back to 1951, the year of the Festival of Britain, when a helicopter was supplied to fly a scheduled service to the site of the Exhibition of which nothing remains today except for the Royal Festival Hall.

From time to time, between the end of the First World War and 1951, the course of the river has, however, been used for serious aeronautical purposes. One of the last occasions before the institution of the Regulations would have been in 1919 when a Fairey IIIC floatplane prototype operated, for a week in May, an experimental newspaper service from Blackfriars to the Isle of Thanet. This enabled readers in Kent to buy their copies of the *Evening News* two hours before their normal sale time.

Perhaps all this actual use of the water as a landing and take-off area owed something to Grahame White who, in the early days at Putney, had regularly flown his Morane floatplane from the water there. Certainly the idea of direct links with other city centres was being tried out in a Vickers Viking III amphibian in 1921 when a pioneer flight had brought the Controller General of Civil Aviation, Major-General Sir Frederick Sykes, from Paris, where the machine rose from the Seine, to the Thames in the shadow of the Houses of Parliament in two hours.

Alan Cobham's DH 50 floatplane lands on the Thames on October 1 1926 following the England-Australia-England flight, which earned him a knighthood (Hawker Siddeley).

Five years later, on October 1, the little blue and silver DH 50 floatplane, which Alan Cobham had brought back from a 28,000 mile flight to Australia, roared low over Westminster Bridge. It landed beneath Big Ben where a launch waited to take him to hear a message of congratulation read by Air Minister Sir Samuel Hoare on behalf of King George V, and later, a knighthood. For this he had received special permission, so did others: the Short Calcutta flying boat, *G-EBVG*, seen at the same spot a little later; the Saunders Roe SRA/1, the first jet flying boat fighter, which arrived spectacularly in more recent times; and a Short Solent which moored in the Pool of London in 1947.

It is only a short step from this part of the river to Waterloo Station and, surprisingly, even this can supply a flying connotation or two. It was here that many a passenger arriving perhaps from their departure points in Hampshire, was surprised to find, suspended from the roof not far from Platform 10, an example of Henri Mignet's £100 *Pouce-du-Ciel*. This was the 'Flying Flea' of the 1930s which was the result of an ill-fated attempt to introduce a cheap aeroplane which anyone could assemble at home on the DIY principle.

A short distance away from the buildings of the terminus there stands an even older, and certainly more surprising, monument to military aviation, one that is seen daily by thousands of passengers who mostly fail to realise that they are looking at a relic from another age. These are the crumbling walls of what was once some Victorian line-side dwellings, last occupied on the night of Saturday, September 29 1917, and destroyed by a pair of bombs from a Staaken 'Giant'

Left *Remains of dwellings destroyed in an air raid during the First World War which may still be seen today.*

Right *August 1928 saw the first Short Calcutta flying boat moored on the Thames for three days during which time it was inspected by several MPs.*

bomber which exploded alongside the railway track at the Waterloo rail entrance.

A number of bombs were scattered over various parts of southern England, both by this machine and the two Gothas which reached London with it, as they met with little in the way of opposition from the defences. Although several hundred rounds were fired by the ground gunners, the shooting was poor and the raiders were fairly immune at the height from which they operated. This part of London is not only a short distance from the river, which the raiders may have used as a guide, for the moon was approaching its maximum on that night, but is also but a brief step from the present site of the Imperial War Museum which has a rich collection of aeronautica from both World Wars.

The manner in which the aircraft collection was begun is not without interest, for the collection on show today owes its establishment to a decision taken as far back as December 1917. This was when the newly-formed Committee of what was, at that time, termed the National War Museum, was informed by its President of the need to preserve specimens of obsolete military aircraft. In the fullness of time the Agricultural Hall at Islington was pressed into service where no less than one hundred machines were ready for exhibition by the end of the following year.

The bulk of these were captured enemy machines as visitors, between 10 am and 8 pm on any day of the week except Tuesdays, Fridays and Sundays,

discovered when they had paid their shilling (5p) entrance fee towards the Royal
Air Force Hospital Fund. (This latter was in Eaton Square with a branch in
Bryanston Square.) The exhibition was opened by Lord Weir on Friday,
November 15 1917, and was to remain open until January 18 1919. In
connection with the show the public school which submitted the greatest
number of descriptive essays of equal merit was offered a prize of *a complete
German aeroplane*, while Boy Scout troops had to be content with an engine
awarded to the winners.

Known German machines on show among the advertisements for Jeyes Fluid
and Thorley's Food included an AEG J-1, Friedrichshafen G III, Fokker Dr I,
Fokker D VII, Pfalz D III, Junkers J-1, Albatros D I and a D Va all in company
with an Ö Aviatik (Berg) and a Halberstadt CL II, while a Pfalz D IIIa was
arranged to portray a 'crash from 6,000 feet' with a tailor's dummy for added
realism in the pilot's seat!

However, the Committee was also responsible for the preservation of Allied
aeroplanes including a Curtiss R2, Spad S VII, Warneford's Morane Type 'L'
parasol monoplane (in which he won the Victoria Cross with the destruction of
Zeppelin *LZ37*), preserved by the Admiralty since 1915, and both a Maurice and
Henri Farman. Alas, all these were destroyed, seemingly because they were not
of British manufacture, together with several others which were. Among the
survivors was the impressive bulk of the only machine to fly at the Battle of
Jutland, the Short 184 seaplane, and this graced the Lambeth Museum until it
was severely damaged by bombing in 1941. It has never returned and the
remains are still in store at Duxford.

Other aircraft in the collection were at several venues including Crystal Palace
and South Kensington and it was here that the Sopwith 2F1 Camel was stored in
1932. This is the machine flown by Lieutenant S.D. Culley to destroy the
Zeppelin *L53* and, when examined some years ago by its former pilot, he

Left *RE8 and Stuart Culley's 2F1 Camel (incorrectly marked* F3043) *as shown in the former War Museum layout.*

Right *The undersurface of Sopwith Camel,* N6812, *which, although now cream, was formerly light blue when it was used to destroy airship* L53.

declared that the undersurface was at one time doped light blue. Subsequent investigation found traces of this shade beneath the fuselage. Unfortunately restorations of this aircraft have not always had a happy result so that, although the spurious serial number at one time applied was later corrected, it still retains wooden centre section struts in place of the metal ones normally associated with the 2F1 version.

Other aircraft of this period of aerial warfare in the Lambeth collection include the Bristol F2B and BE2C, plus, of course, the Zeppelin observation car already mentioned in another chapter and all these were at one time on show in the old Air Services Room on the first floor which now does duty as the Museum Cinema. The only landplane displayed in another part was the RE8, suspended over the guns in the principal Army Gallery, but in more recent times this has been lost to Duxford. The end of the Second World War found the necessity to augment the collection of aircraft and the new additions included a Heinkel He 162A, Focke-Wulf 190A-8, de Havilland Mosquito, Spitfire, Messerschmitt 163, part of a Typhoon and a Zero-Sen, the forward fuselage of an Avro Lancaster and a Fieseler Fi 103 flying bomb, while at one time there was a Gloster Meteor. There is a Swordfish upstairs.

Strangely, Victoria Embankment has come to have a multitude of aeronautical associations and the nearest one to the visitor approaching from the direction of Westminster Bridge is the statue of Hugh Trenchard, the 'father of the Royal Air Force'. This was unveiled by the Prime Minister, Harold Macmillan, and the choice of site in Victoria Embankment Gardens was plainly due to the short distance which separates it from the Royal Air Force Memorial and because the windows of the present Ministry of Defence (Air) buildings look down on the little park.

It would seem that here is a suitable point in our narrative to record a little-known anecdote about this officer. It was the opening months of 1918 that, at

Left *Lord Trenchard's statue in Victoria Embankment Gardens.*
Right *Oscar Nemon's statue of Viscount Portal of Hungerford in the same park shows a different style of sculpture.*

No 40 (Training) Squadron, Captain Taylor, a superb pilot but a fanatical perfectionist, was putting his trembling fledglings through their paces with the usual flow of abuse and invective. At this particular moment on the cold February morning the instructor seemed to be so absorbed in this, that his attention was not attracted by the sight of the strange Avro machine coming in to land. The first attempt to do so, was much too high, so that the pilot went round again and, although successful on this attempt, the result was something short of perfect.

The machine was taxied up to a hangar and just as the pilot was climbing down the redoubtable Captain sauntered over and began to upbraid the officer, unmindful, that his open trench coat revealed underneath a tunic adorned with the red tabs of a Staff Officer. In the pregnant moment of silence when the new arrival unwound his scarf, Taylor looked him up and down before speaking. 'Where do you think you learned to fly?', he sneered, 'Fancy coming to a place where I am trying to teach. What sort of example do you think you are?', adding as a final jibe, 'A Staff Officer should be able to do better than that!'. While this was going on, the victim had taken off his goggles and was now struggling with his heavy coat.

'Well Captain', he replied, 'I'm sorry I put on such a bad show. I'm rather a poor pilot, I'm afraid, as I'm at a desk most of the time and very seldom get a chance to fly. This is the first time I've had an Avro up but I won't come back and give another bad exhibition after this! I've heard all about you and that you're a good instructor and how you tick people off if they don't show up properly.' By now the speaker had rid himself of his wool hat and trench coat so that the assembled pupils now saw him to be a man of strong features and prominent brows in a service jacket liberally adorned with rank braid, several rows of medal ribbons and a General's insignia. It was none other than 'Boom' Trenchard himself, commander of the entire Flying Corps! For once Taylor was lost for words and scarcely seemed to notice the wink that the General delivered in the direction of the pupils as he patted their instructor on the shoulder and, with a twinkle in his eye remarked 'Come on Captain, don't be ashamed of having done a good job'.

A few yards away from the monument to this illustrious officer, stands the statue of another, Marshal of the Royal Air Force, Viscount Portal of Hungerford. This piece of sculpture is the work of Oscar Nemon and was unveiled, once more, by Harold Macmillan on May 21 1975. The choice of this little garden for these effigies, which they share with the statue of General Gordon, is undoubtedly because of the proximity of Sir Reginald Blomfield's Royal Air Force Memorial. This stands on the Victoria Embankment proper, surmounted by the massive bronze eagle designed by Conrad Parlanti who, in later years, was also responsible for the 'Victory Bells' marketed after 'VE-day' by The Royal Air Force Benevolent Fund and cast from the metal salvaged from Nazi aircraft.

When the memorial was unveiled by Edward, Prince of Wales, accompanied by the Duke of York in July 1923, many remarked on the fact that the position chosen for the 54 foot-high column at the top of Whitehall stairs presented the eagle facing the river. This position was chosen as it gives the idea of unfettered flight over the open space of the water. The additonal inscription commemorating the fallen of the Second World War was unveiled on September 15 1946 by Lord Trenchard, then a 74-year-old Marshal of the Royal Air Force.

A short distance up the connecting Horse Guards Avenue brings one into Whitehall and an area which is particularly rich in aviation associations. Mention has already been made of the pile of the MOD (Air) with its famous 'pillared hall', scene of occasional exhibitions but few know the significance of the 18th century building to the south which occupies the site next but one to the Banqueting Hall. This is Gwydyr House which serves as the Welsh Office but, before the erection of Adastral House in Kingsway, this was the Air Ministry. The name is taken from that of Sir Peter Burrell who became Lord Gwydyr (the title being that of property in Carnarvonshire) and the house was completed in 1796. Its erection had caused some problems, for the proximity of the site to the river meant that since there was a distinct danger of the Thames bursting its banks at this point and flooding the area, a fact that has an almost modern ring. It was therefore necessary for considerable pile-driving to be carried out during the laying of the foundations.

The actual builder was one John Marquand, a surveyor of Woods and Forests in the government office of that name and soon after, the house was bought by the government for use by various departments including the old Poor Law Board. Until 1886 it presented a slightly different appearance for it was not until then that the top storey was added but following that date this historic building has undergone few significant external changes.

Not quite directly facing this on the other side of Whitehall stands Horse Guards Parade and, probably due to the open area that is associated with the ceremonies held in connection with the monarch's Birthday Parade, it may not at first seem to have much to claim in the way of aerial atmosphere.

However, this is by no means correct and probably the first of these was that in connection with the formation of the London Defences of September 1915. These were to be commanded by Admiral Sir Percy Scott to whom the First Lord of the Admiralty Mr (later Lord) Balfour introduced Lieutenant Commander Rawlinson, RNVR, who was to assist in the establishment of the defences based on the pattern of those serving Paris, of which he had some experience.

This meeting took place in the First Lord's room overlooking the Parade and from here the junior officer was directed to investigate the position as to the availability of a suitable French anti-aircraft weapon. The night boat from Dover ensured that he was at the Paris Arsenal the following morning and the authorisation for a mobile gun to be procured was shortly afterwards granted by the Commander-in-Chief, General Joffre.

Once the formalities were completed, the gun was driven at high speed for the coast, with Rawlinson finally overtaking it on the road to Boulogne at about 5 pm on the following day. He was able to superintend its loading a little later, thus ensuring the arrival of the new weapon in the harbour at Newhaven a little before 7 am on the next morning. Thus it was that the *Canon Automobile*, cleaned and ready for inspection, was, at 2.30 pm on the same afternoon, drawn up on Horse Guards Parade under the First Lord's window. The efficiency and speed of the whole operation is proved by the fact which later emerged that, while the gun and crew was actually drawn up and awaiting the pleasure of Mr Balfour, the signal from the Admiralty requesting the weapon was about to be drafted!

Less efficient was the train of events which, in the very different world of only 27 years later, was to bring the Bucknell Enquiry, personally ordered by Winston Churchill, to a small room overlooking this same area. The Board met to look into the sorry fiasco which attended the escape up the English Channel of the three Nazi warships, *Scharnhorst, Gneisenau* and *Prinz Eugen*, not that this was in any measure an exclusively Air Force affair, although it had been initiated by two pilots operating from an aerodrome close to London—Kenley in Surrey.

Perhaps the strongest aviation connection with Horse Guards Parade is the succession of exhibitions of military aircraft which have been held here from time to time since 1946. It was at this first show that some attention was attracted by what at the time was termed 'an early Hurricane', none other, in fact, than *L1592*, the specimen which is now on view in South Kensington's Science Museum. An interesting fact is that, when it was shown in connection with the Victory Celebrations, the finish was the conventional one of the day, with 'Sky' fuselage band and spinner and the single identification 'I' behind the

Above left *One of a pair of RAF badges that flank the entrance of the main Ministry of Defence building in Whitehall.*
Above right *The great bronze eagle that surmounts the Royal Air Force Memorial at the top of Whitehall stairs.*
Below *Gwydyr House, the old Air Ministry building, now the Welsh Office.*

DT A

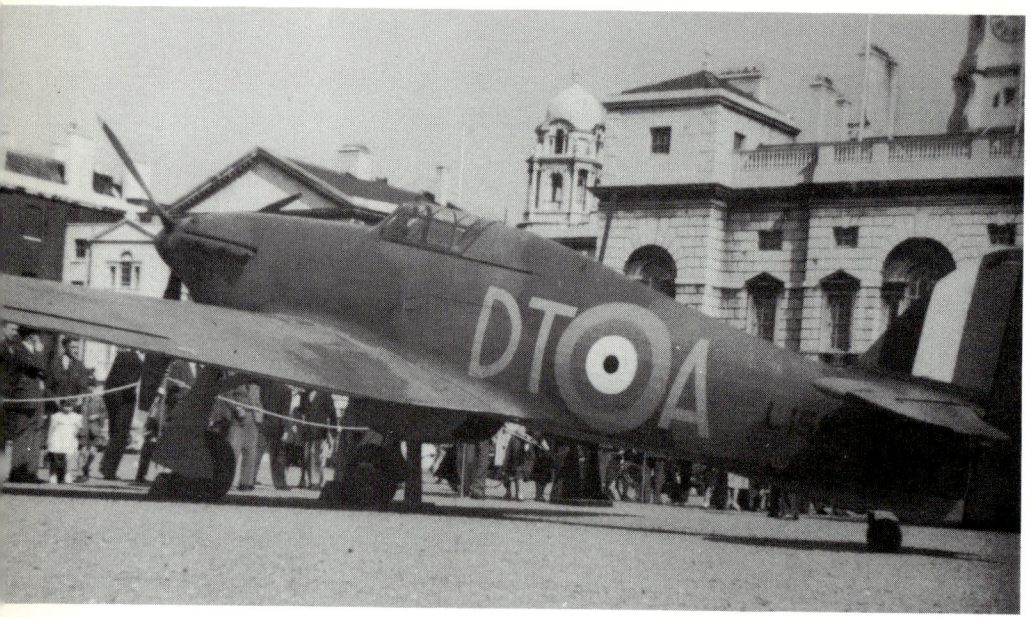

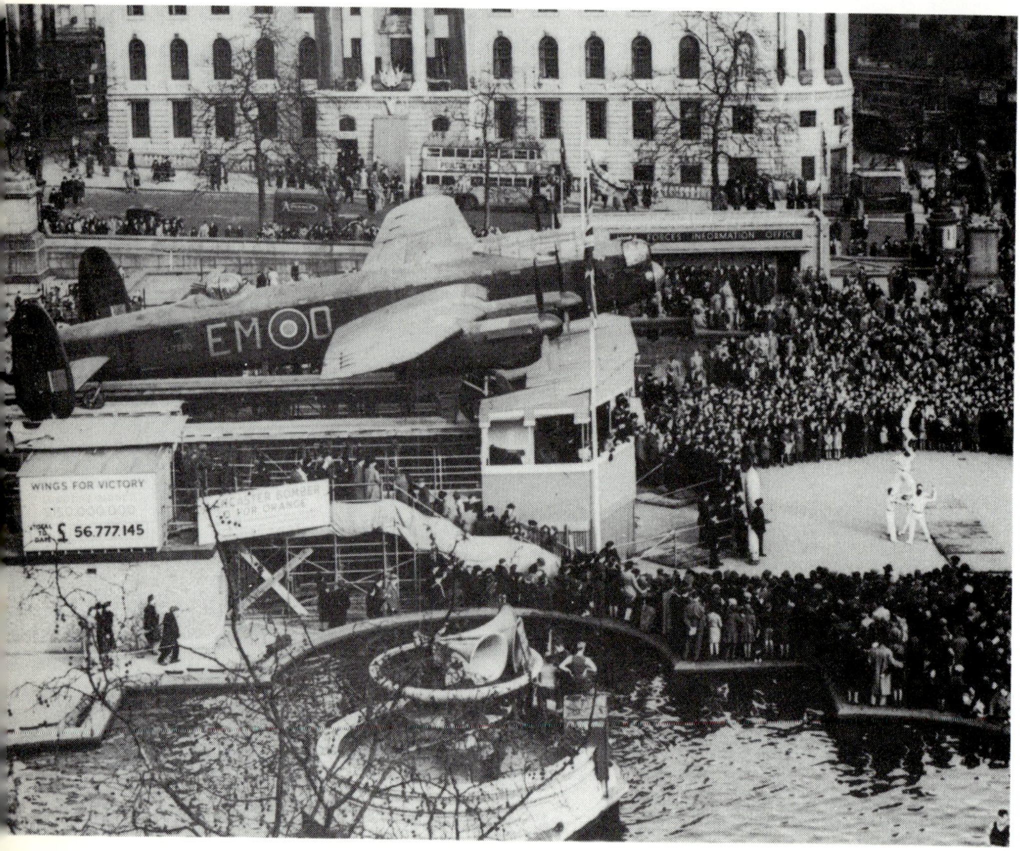
FORCES INFORMATION OFFICE
EM D
WINGS FOR VICTORY
£ 56.777.145

Left *Hawker Hurricane, L1592, now displayed in the Science Museum, is seen here in markings carried for display purposes in 1950.*

Below left *Avro Lancaster, L7580, on display in wartime London's Trafalgar Square in connection with War Savings* (Imperial War Museum).

Right *Spitfire, K9942, used in 1960 to advertise a Battle of Britain exhibition in what is now the main Ministry of Defence building.*

roundel. It was back again for another Horse Guards Parade display in 1950, with a new identity, DT-A, and, as before, the collection included a wide variety of RAF and ex-enemy machines. Ten years later, it was a case of 'the mixture as before' marking the 20th anniversary of the Battle of Britain and, with a Spitfire on a trailer by way of publicity, the Ministry of Defence building in Whitehall presented a large collection of models to demonstrate the evolution of the military aeroplane.

Perhaps the culmination of all these open air shows during the summer months took place in 1968 in connection with the celebrations of the RAF's Golden Jubilee. By now, the larger part of the collection was found by the then-unopened RAF Museum at Hendon, although many of the old favourites had acquired new looks, including Gloster Gladiator *K8042,* now no longer in drab camouflage but resplendent in the finish of the 1930s, and the veteran Sopwith F1 Camel among others.

It would be wrong to give the impression that this show dealt with the past, however, for an attempt to give as complete a history as possible of RAF heritage was aimed for, *XM479* represented Hunting Jet Provost T3 trainers of the day, while Supermarine S6A, *N248,* now in the safekeeping of the R.J. Mitchel Hall Aircraft Museum at Southampton, masqueraded as the S6B *S1596* in the marking that had first been applied as long ago as 1942 when it had been used to represent this machine in the wartime film about the life and work of Mitchell, *The First of the Few* starring Leslie Howard, David Niven and Rosamund John. The very first public appearance in this finish had been at Eastleigh, Southampton, in 1946 when, during an air display, it had been towed on a trailer behind a tractor round the public enclosures.

Only a short distance from the scene of this activity lies Trafalgar Square and even here linger the last remnants of a past atmosphere of aviation. The occasion was one of the many war-savings drives of the years of conflict, this

A barrage balloon was used to suspend a banner showing the London War Savings total (Imperial War Museum).

one bearing the succinct title of 'Wings for Victory'. The focal point was provided by the exhibition of a complete Avro Lancaster Mk I, *L7580*, making it quite an early model since it had been delivered at the very end of a batch of 200 machines from which the first 157 had been Avro Manchesters. The markings declared it as having served with No 207 Squadron which had operated the type since March 1942 so that it was marked *EM0* and was declared as having completed 13 operational flights over Nazi Germany. It was displayed, between two of the fountains beloved by New Year revellers, on a type of dais which prevented close examination and competed for the honours as a 'crowd puller' with a handsome coach and four which arrived to excite the crowds on certain mornings!

From this point is is possible to make one's way with ease to Birdcage Walk to the south of which, at the Parliament Square end, lies Queen Anne's Gate. During the late summer of 1914 this was the London Home of the Honourable Maurice Baring who recalled that just after the outbreak of the First World War, on Sunday, August 9, 'various people came to No 32'. The reason for this stream of callers was that Major Baring had been appointed ADC to Hugh Trenchard and the Royal Flying Corps was about to leave for France. Long years after, the latter was to write of his old assistant that, 'there never was a staff officer in any country, in any nation, in any century, like Major Maurice Baring. He was the most unselfish man I have ever met or am likely to met . . . words fail me in describing this man'.

Long before the RFC appointment had been made, Baring, the fifth son of Baron Revelstoke, had left Eton for Trinity College, Cambridge, already with a

One of the two Sphinx figures at the foot of Cleopatra's Needle still bears signs of damage sustained in an air raid in August 1917.

reputation for being a wit, and in the fullness of time he became fluent in no less than seven languages. With a skill such as this it was almost inevitable that he should join the staff of *The Morning Post* as War Correspondent covering the Russo-Japanese conflict of 1904 before becoming *The Times* correspondent in the Balkans eight years later. In the years to come he was deservedly to earn a name for himself as a man of letters, author, poet, playwright and essayist as well as excelling in the Diplomatic Service. Many of his earlier poems were concerned with the first days on foreign soil of the new flying service and it is his lengthy work, *Per Ardua*, which immortalised the departure of Nos 2, 3, 4 and 5 Squadrons of the 1914 British Expeditionary Force when he describes with poetic licence the '. . . Squadrons three of the RFC' which 'Flew over the sea from England to France, as gaily as to a dance'.

It is no great distance from Queen Anne's Gate, back to Victoria Embankment. If this is done via the Strand and perhaps Adam Street one arrives in the northern part of the twin gardens where the statues already discussed, are situated. In front stands the obilisk of Cleopatra's Needle and even this bears the stamp of past air activity. It does not call for a very detailed examination of the two Sphinx figures which guard the ancient Egyptian monument to discover that the more northerly one has a moderate sized puncture in its paw. The manner in which this came about provides another interesting link with the early days of total warfare.

There had been several air raids by aeroplanes over, not only London, but Kent and Essex too, during the summer of 1917 and that made at night on August 22 was the first of a new phase in the attacks as they were carried out by moonlight. The later hours of darkness on the evening of September 3 had seen a number of Gotha bombers operating from bases in Belgium, and four machines of *Kagohl 3*, ie, *Kampfgeschwader 3* of the *Oberste Heeresleitung* or OHL were seen in the bright 'bombers' moon to cross the coast of Kent.

The result of their night's work was the most serious incident of First World War bombing, for a group of 50 kg bombs released over Chatham demolished a drill hall where a large number of sailors were billeted, killing over 130 of them and injuring almost as many others. With this in mind, those who were equipped with a knowledge of these happenings faced the sound of the warning maroons on the following night with a feeling of impending doom. It was quickly evident that the seven bombers heading over the coasts of Sussex, Essex and Kent were intent on the capital as their target, although three of this force turned back, one each over Dover and Margate while another described a circle north of Bawdsey before setting course again over the sea.

It was not long before the night was rent by the sound of bombs coming down, the Londoners looked at each other with unspoken questions on their faces, in silent enquiry if these were the 'big' bombs of the day, the same as had caused so much havoc in the Naval dockyard the previous night, the 'Crashing Christophers'. Despite the small number of raiders they did not all arrive over London at one and the same time, with the result that it was two hours before the Boy Scout bugles sounded the welcome notes of the 'All Clear'. The necessity of keeping a warning operating for such a lengthy period was regarded as a particularly barbarous refinement at the time.

The bombs which dropped fell mainly in the vicinity of Oxford Street but a few exploded along the Thames. One did so near to Charing Cross Hospital, doing little damage save for blowing in the windows, while another scored a near miss to the Hotel Cecil which flanked Cleopatra's Needle and was then in use as a RNAS headquarters in which capacity it enjoyed the name of 'The Bolo' in the Service.

However, while the building suffered minor damage from the blast, a tram passing along the Victoria Embankment caught the full effect, and while several passengers were killed outright, Mr Alfred Buckle, the driver, despite one leg being almost severed by a piece of shrapnel, managed to bring his vehicle to a stop. Other splinters damaged the Egyptian obelisk and although there is now little sign of the damage which was done to the stonework, that done to the bronze figures remains to this day, mute testimony to the night when 800 shells were flung skywards at the raiders without effect by the anti-aircraft batteries. Unfortunately the 32-year-old tram driver died before he could be admitted to hospital, despite the attempts to stop the bleeding by a pair of American Army doctors who were at the scene.

This particular tour of London's aviation associations to be found near the river has, by now, come almost full cycle and the reader is presented with alternatives as to which course he chooses; both present some additional points of aeronautical interest. The first possibility is that one retraces one's steps over Westminster Bridge, perhaps to Waterloo Station, while the second indicates a journey round Broad Sanctuary, possibly *en route* for Victoria Street.

Those who cross the river are approaching the district of Newington Butts and it was in this area that Vincent Lunardi made something of a come-back in Georgian London, following his rough treatment by the crowd on the occasion of his premature landing in Tottenham Court Road. As a gesture of goodwill,

Ballooning was still popular at the beginning of the present century although gas had replaced hot air as a lifting agent completely. This photograph dates from about 1908 and shows The Hon C.S. Rolls among the passengers rising from Battersea. The idea of the Royal Aero Club is said to have been formulated during a journey similar to this.

Lunardi announced that the new attempt would be presented without any charge to the spectators and tickets were issued permitting them to witness the actual inflation of the envelope with hydrogen at the Royal George Rotunda in St George's Fields. Not unnaturally the gathering was one of huge proportions, for 1,000,000 people assembled to see the ascent.

The chosen day was June 29 1785 and the preparations went ahead without interruption but, towards the end of these, it became evident once more that the lift of the balloon had been over-estimated and it would be incapable of taking the three persons hoped for, namely Lunardi, George Biggin and Letetia Sage. In retrospect it is perhaps not too ungallant to suggest that this lovely creature was something of a problem to voyagers in lighter-than-air-craft for she was of

Sir Douglas Bader, CBE, DSO, DFC, pauses outside Westminster Abbey following an RAF Jubilee Service in 1978.

splendid Junoesque proportions, tipping the scales at 14 stone.

When it became clear that three passengers were out of the question, Lunardi immediately waived his right to captain the vessel and the final ascent was made with Biggin and Mrs Sage, the latter thus becoming, as she liked to describe herself thereafter, 'the first English female aerial traveller'.

As with any other balloon passengers, the pair were no more than the playthings of the wind and the currents now wafted them across the river in the direction of Westminster and a short time later they found themselves looking down on Piccadilly, a fact that we know now, thanks to the care with which the lady on board noted the details in a small book which she had taken for this express purpose and into which she noted every detail of the new experience with 'complete composure'.

During this time George Biggin, alone and in control of a balloon for the first time, found himself better able to control his craft than he had imagined and later, when the vessel had been wafted in a new direction that took it across the river in the direction of Harrow, he found no difficulty in bringing it to land once more in that area, after having been aloft for about an hour. Alas, posterity has left us no accurate pictorial record of the flight beneath the giant· envelope with its Union Flag design, for all show three passengers not the two of reality: an expression of hope rather than fact.

Should one not return to the vicinity of the St George's Fields, for which search may be made in vain today, but rather via the area of Parliament Square, it is necessary to pass that part of the Embankment on the other side of the Thames which marked a new departure for 10 days in May of 1973. It was then, under the sponsorship of *The Daily Express*, that there was presented on the South Bank what at the time was termed The First London Plane Show and along the side of the river was displayed not only light business and executive aeroplanes but also helicopters, aviation equipment and services plus the many aspects of modern flying training. The venue was entirely appropriate, for the same vicinity had seen (during the Festival of Britain of 1951 in the Transport Pavilion) such historic machines on display as the Gloster E28/9 jet (which remains in London to this day in the Aeronautical Gallery of the Science Museum) and the de Havilland DH 88 Comet, *Grosvenor House*, which had been the winner of the Mildenhall to Melbourne Air Race and the £15,000 prize given by Sir MacPherson Robertson in 1934. This design not only foreshadowed the Mosquito of war-time memory but also may still have flying hours before it, some half a century after the event in the care of the Shuttleworth Trust, Bedfordshire.

The shadow of Big Ben itself had also seen flying connections of a different nature as the Abbey interior is lit by the great Battle of Britain Memorial Window, unveiled by His Majesty King George VI on July 10 1947. In more recent times the same sacred fabric saw the Service of Thanksgiving to commemorate the 60th Anniversary of the formation of the Royal Air Force at 11 am on Saturday April 1 1948. The congregation which includes a vast number of Service and ex-Service men and women of every rank and station was led by Her Majesty Queen Elizabeth with Prince Philip.

When the Lord Mayor's Show of 1926 presented a wingless G-EBFO, now

A Sabena S-55 helicopter lands near County Hall, Westminster (Sabena).

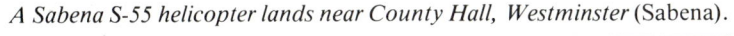

fitted with a wheeled undercarriage and alongside a notice describing its journey not only to Australia and back but also the round trips to Rangoon and Capetown, it seemed to many that the art of flying had reached its zenith. With the passing of the subsequent golden years of pioneer aviation in the 1920s it certainly seemed as if this was so. Then, in 1969 there was something of an echo of the foolish, happy days of four decades earlier. The event was the 1969 trans-Atlantic air race which was organised by the *Daily Mail* in commemoration of the first non-stop crossing of the 'herring pond' in 1919, eight years before Lindbergh and his twin-motor machine which only offered twice as much to go wrong.

The *Daily Mail* Race of 1969 was in no way tied to the river, but, as has already been explained, London's waterway is the highway for helicopters, and since the race expected to be run from the Post Office Tower to New York's Empire State Building, rotary wing machines were a first choice by some for the London end of the course. Typical was an Army entry using an Air Corps Scout helicopter from which, as it hovered only feet above the surface of 'Father Thames', Cadet Wynne-Davies was seen to execute a spectacular leap on to the deck of a waiting launch as part of the wild dash to the next rendezvous point.

There were others: and for two, an 87 foot barge of the Mercantile Lighterage Company, named *Wayherd*, served as a helipad. Moored in the normal manner she was too high in the water for competitors to climb aboard from their inflatable boats, so that a carefully graded load had to be put aboard to bring the barge to the correct level. One of the contestants using this vessel was none other than Stirling Moss, forsaking the road for the air on this occasion, and also Clement Freud in the company of 10-year-old Dominic who also found themselves on the Thames as a stepping-stone along their route to the Aer Lingus machine which was to whisk them across the Atlantic Ocean.

Chapter 4

Among the stores

There are, basically, two types of aviation connection which a centre of population can have, those which may be termed 'direct' in that a venue has some immediate connection with an event, either brief or of long standing, or there is the alternative one which is of an indirect nature and for the most part consists of unwelcome attentions from the air, either in the form of an aeroplane coming to grief there or the results of a bombing attack.

Naturally the former has pride of place in a book of this nature and it is felt that the second category more properly belongs to a work on disasters so there are few between these covers. The last type has some claim to inclusion since quite often the results of a war-like action has altered the whole course of history in the immediate area, a fact which also explains why the majority of this type of incident, as described here, took place during the First World War. This is for two reasons, firstly because the extent of bombing by the Nazi Luftwaffe was too vast to cover, along with other material, and also due to the fact that incidents during the years of 1914 to 1918 are fast passing out of personal recollection and into the limbo of history so that it is felt that before it is too late at least some representative ones should be set down.

As a starting point, the eastern end of the Strand is unique in having the Royal Air Force Church, not to be confused with the neighbouring St Mary-le-Strand. The original fabric was designed by Sir Christopher Wren, and was opened for worship in 1682. It was used by generations of Londoners thereafter, including the children who would each be presented annually with an orange and a lemon at the conclusion of their personal service, as indeed they still are, in commemoration of the church of St Clement Danes in the nursery rhyme.

At the height of the London 'blitz' all this came to an abrupt end when the building was gutted and the peal of bells with their ancient message were seemingly silenced for ever when they crashed down to destruction at the foot of the steeple. Only one event marked the clearance of the worst of the rubble when workers were surprised to find five small coins on what was left of a pew. At first they were throught to be sixpenny pieces but the colour proved to be only that of the fine layer of ash which covered them; they were in fact farthings, perhaps an echo of another line of the jingle brought by a long-forgotten child of the 1930s, pushed under a cushion and lost. In the years that followed there was nothing remarkable to bring attention back to the silent shell in the middle of the Strand except the humourist who placed the finial from a

Left *Wren's St Clement Dane's in the Strand as it appears today.*

Right *Heston Phoenix G-AEVS, used to test airborne radio equipment by the Standard Telephones and Cables Company during the 1930s* (STC).

large ornamental railing on the head of the statue of Doctor Johnson behind the building to serve him as a hat!

Sunday, October 19 1958, saw life return permanently to St Clements, however, for on that day the church, now rebuilt and with features added that Wren had originally wished for but failed to achieve, was formally reconsecrated in the presence of Queen Elizabeth II and the Duke of Edinburgh. The recast bells, plus the addition of a new eleventh one, rang out again over London for the first time from a church designated for the Royal Air Force. Inside, the floor is covered with 735 badges which have become so famous but even now are showing unfortunate signs of wear. The memorials include those varying from that marking the resting place of Lord Tedder's ashes, brought home here by an admiring retired RAF Sergeant, to a poignant one dedicated to Thelma Bader.

Not far away is the site of the former Gaiety Theatre, the mecca of men from all three branches of the armed forces in 1914-18 and, in the way of things, a surprising number of ex-airmen of the period who survived have entered the ranks of show-biz: Ben Travers, Jack Warner, Jack Payne, Billy Cotton, Rex Palmer, Roland Culver and Ronald Adam to name but a few, and it is to the latter that the author is grateful for recalling the night when Leefe Robinson, hero of the destruction of the airship, *SL11*, entered a box here. 'He was every girl's epitome of the chocolate-box hero. The whole house stood and applauded for fully five minutes'.

On the other side of the Strand are the offices of Standard Telephones and Cables Limited and this organisation, while resident at Connaught House,

Aldwych, not only participated directly in aviation but also made important contributions to the British defence system of the day, so that it is not surprising that during the 1930s a Heston Phoenix monoplane, *G-AEVS*, was purchased and, with the name *Standard Radio* on the nose, flew regularly to test airborne radio equipment.

The company established itself in the aircraft communication field in 1931 with a range of HF and MF equipment and this was followed by a Standard Beam Approach System which was to become famous throughout the world while, five years later, there appeared the revolutionary R10-A Fighter Radio designed to overcome the problems associated with fast single-seat machines. For the development work of this pioneer equipment, regarded at the time as being the smallest in the world, a full-time pilot was employed, H.M. Samuelson. Another field in which STC made great strides was that of instrument landing systems. Croydon, Gatwick, Heston and Collinstown were all thus equipped before 1937, the year when the company's 'Lorenz' Blind Approach System was adopted, the forerunner of war-time Bomber Command's landing aid and the widely-used SBA found at international airports to this day.

A glance through the aeronautical directories of all periods, particularly the years of conflict, will quickly show a vast number of aviation addresses for the streets which lie north of the Strand, but for the most part these were registered offices. In passing them on the way to Piccadilly Circus it is of interest to note that in the early years of civil air transportation, the offices of Air Union could be found in the Haymarket. These were forerunners of the modern Air France offices at No 52, a thoroughfare it shared with Imperial Airways in Charles Street and Redline Motor Spirit at No 16.

At first it would seem that the hub, of which Eros is the centre, has stood for many ages without change. This is not so for some reconstruction was untertaken in 1923 which included Regent Street after John Nash's dignified architecture fell victim to the tide of ferro-concrete. It was only a very short time after this that the building of Swan & Edgar was demolished. Incidentally, in those days, Eros pointed to this although it faces a different direction today.

The building which came down had almost been demolished violently in 1917 during an attack on the night of October 19/20, known as the 'Silent Raid'. It is historically significant since it constituted the last-ever assault on London by an airship. The bomb which fell, recorded in contemporary accounts as being of 2 cwt capacity, was the only one to fall on the centre of the capital that evening and the manner in which it was brought is somewhat unusual. No less than 11 airships had set out in the dark of the late autumn evening and all crossed the British coast, either north of the Wash or over East Anglia. L45, the vessel destined for the historic role, made landfall at 8.20 pm beyond Grimsby and, perhaps because of the gale and extreme cold experienced at 16,000 feet, proceeded to wander about in the area before making off towards Louth. From this point a course was resolutely set in the direction of Leicester but, before the city came into view, bearings again became confused and, after a short detour, a new course brought L45 directly over Northampton. Here several bombs were dropped before the vessel set off in a more or less southerly direction so that it passed over Leighton Buzzard and within sight of St Albans.

The choice of direction was to some extent involuntary, for the airship was in fact running before the gale, a wind that those on the ground knew nothing of and it was not long before the first signs of a large river attracted the attention of the look-out. That this was the Thames there was no question but the precise position of L45 was quickly established by the clusters of lights below for, although they were dimmed, there was no complete blackout and the glow at once confirmed that they were over some large city. The two factors of the river and the lights were soon after confirmed as marking the capital as the helmsman was, a little later, able to pick out, although indistinctly, 'two great railway stations', probably St Pancras and King's Cross, although little else was at all clear.

That they had blundered across London momentarily raised the spirits of the crew for they were troubled by the gale driving them so far into what was known to be a well-defended area, although as yet no guns fired and no lights probed the sky; while at the same time all on board were suffering agonies from the cold. With such a target below, the final bomb in the racks was released and, all the excitement over in a moment, the great dirigible was again rushing onward through the darkness. The bomb exploded in the roadway adjacent to the frontage of Swan & Edgar's store in Piccadilly, penetrating to a depth where it exposed the gas mains and telephone cables. Loss of life also was considerable, there being 18 injuries and seven dead. It was the only bomb to fall on central London that night and thereafter attacks on the capital were to be made by aeroplanes.

Despite the cautious attitude of the newspapers which marked the reporting of the day, there was no hiding the fact that a mixture of luck and skill had enabled the Zeppelin fleet to deliver a blow where it would not swiftly be

Part of the hole in the road outside Swan and Edgar's shop in Piccadilly (Imperial War Museum).

forgotten. It was not long before those who had seen the site of the incident spread the details of the gaping, 12 foot hole in the roadway. *L45* meanwhile escaped over Chatham and Hastings.

The actor, Charles Hawtry, a member of the gifted family which still graces the theatrical world to this day, was, at the moment of the explosion, on the point of entering a nearby restaurant and suffered some bruising and shock when the blast flung him through the doorway into the building. By now, Londoners were becoming nervous at the frequency of attacks such as this and the Underground stations had for some time been pressed into service as shelters, thus establishing a precedent which was to be followed on a much greater scale in the 'Blitz' of nearly a quarter of a century later. Only a few nights before the Piccadilly bombing there had been a grim incident when a number of people had been killed in the panic when the explosion of the warning maroons had been mistaken for those of bombs from undetected raiders.

Not far away between Berkeley Square and New Bond Street runs Bruton Street and it was here, at No 13, that the Royal Flying Corps Club was established in January 1917. The founder was Lieutenant Colonel W.C. Bersey and he was to see the name of the new centre undergo a change to that which it still enjoys when, in November the following year, it became the Royal Air

Force Club. The original buildings did duty until the closure of the Club at the end of the same year when the endowment by the founder expired. The present magnificent buildings were taken over in January 1922 and 128 Piccadilly became the world-famous mecca of flying men and women not only of the RAF and Commonwealth but also of the world.

The Royal Air Force Club, as we know it today, was the result of a gift of a total £350,000 by one of the former Presidents of the old Air Board, Lord Cowdray. He wished, in this practical way, to create a memorial to his son who had been killed on the Western Front during the Great War, as it was then termed. At the same time he hoped to pay tribute to what he termed the 'brilliant and superlatively heroic work' of the airmen in that conflict. Lord Cowdray is also remembered for his work which contributed to the formation of a separate Air Ministry which today survives in the Ministry of Defence (Air).

At this point we should, in effect, not allow our enthusiasm for this fine organisation (which overlooks Green Park and also has its aviation associations, as we shall see) divert our attention too far from the eastern end of Piccadilly. Across the road from Bond Street at Fortnum & Masons is another item of forgotten aeronautica. The present store dates from a re-building programme undertaken between 1923 and 1925 although the firm is, of course, much older, dating from the first half of the 18th century, so that the surrounding area had plenty of time to absorb the effects of the widening business.

One of the early enterprises in the day and age of the horse as the only means

Left *The former RFC (now RAF) Club building in Bruton Street.*

Above right *The RAF Club building at 128, Piccadilly which resulted from a gift of £350,000 in 1922.*

Right *One of the entrances to Mason's Yard in Duke Street, St James's.*

An SE5a fighter converted to become a contestant in the Oxford Team for the University Air Race of 1921.

of land transport was the establishment of Mason's Stables and this took place in the mews yard on the left hand side of Duke Street, St James's, a turning a little west of the store. That this should rapidly become known as Mason's Yard was only part of the natural order of things and to this day the name may be seen on a plate at each side of an excellent and ancient public house known as 'The Chequers'.

At first it may seem that here is little to connect this yard with the art of flying but the change was to take place during the First World War by which time part of the hidden square had been converted for use as warehouses. It was to a room here (the Royal Flying Corps Despatch Office) that officers were directed to receive their orders to be sent to the Front. Perhaps at this point we may stretch out geographical connections a little and recall that many of these pilots engaged in ferry duties would also have been familiar with Portman Square, off Baker Street, where the Central Ferry Pilots' Pool office was in 1918.

The flyers of that age who survived would, in succeeding years, have been familiar with the area for other reasons as it is only a short distance away that the Criterion Restaurant was, in the mid-1930s, to become one of the accepted centres for reunions. It is interesting to note that, at a typical reunion organised on the night of Saturday March 2 1935 by 55 Squadron RAF, the price of a dinner to members was eight shillings and sixpence (42½p), while guests might be introduced for an extra ten and six (52½p).

By this time aviation had tended to become more widely accepted, perhaps due to the healing effect of time on the memory of the war years, although there were small reminders still scattered about for all to see. One may still be noticed by those who seek it out at the Royal Academy. Fairly high up on one of the walls there is a small plaque recording that: 'This, No.9 Gallery was wrecked in an air raid in the Great War. September 24th, 1917'. The attack in question had been the first for some three weeks and had resulted in some heavy damage by the standards of the period. Bombs had fallen in Southampton Row, outside the Bedford Hotel there, near to St Thomas' Hospital, and in the River Thames in addition to Green Park, not far from the Ritz and further afield in Dean's Yard. Gothas were the raiders on this occasion and the largest bombs dropped were 50 kg.

However, as already stated, these unhappy events had passed into the middle distance of some Londoners' memories after the passage of a little over a decade and the airman (who had been looked upon as crank, hero and something akin to a criminal all in the space of the lifespan of many) was now becoming at least accepted. Perhaps part of the contribution to this change of attitude had been the gradual introduction of flying as a sport, and one such event had been organised as early as July 1921 when a University Air Race, Oxford versus Cambridge, had been flown in surplus wartime SE5A fighters.

Another such event which had helped to popularise sports flying had been the King's Cup, a competition sponsored by the presentation of an annual Trophy by King George V. It had been the securing of this award in 1930 which had brought to London a unique flying event, for the winner that year had been a girl, the first time that this had been done. She was Miss Winifred S. Brown and the aircraft which she had flown was an Avro Avian III, *G-EBVZ,* with a winning speed of 103 mph. Had a male pilot been that year's winner nothing more would probably have been made of the matter but, unfortunately, the young Manchester girl fell victim to certain exploiters, with the result that she agreed to appear on the London variety stage. Thus it was that the new decade opened with something 'completely different' when Winifred Brown appeared nightly for a season on the stage of the Coliseum, clad in flying kit and beside her winning aircraft. Unfortunately, although there were certainly those theatre-goers who welcomed the chance to acclaim their heroine in this manner, the affair was looked on with a measure of disgust in the aviation world so that the whole became a matter for ridicule.

Even so, there have been some very happy marriages of the theatrical world and that of aviation. One took place only three years later when the autumn of 1933 found the Lyric Theatre presenting *The Ace*, a play set in a war-time German *staffel*. The star was Raymond Massey who portrayed Kurt von Hagen, the fictional victor of 49 combats. Yet such entertainment is rare and it is not until 1981 that something similar appeared on the London stage when June saw the curtain rise on John Gray's *Billy Bishop Goes to War* at the Comedy Theatre, Panton Street. This was a musical though, with Eric Peterson in the title role, which had been originally produced in Vancouver.

Other aviation connections, both major and minor, may be found in abundance throughout the area and some have left a mark which may still be seen, others exist only now in the works of historians. Of the former there is the small representation of the RAF badge which surmounts the Roll of Honour over the staircase at Liberty's store, while among the unmarked flying connections must be placed the balloon ascents from Green Park. The most important of these had been part of the celebrations of King George IV's coronation, and it will be remembered that it was this monarch who, as Prince Regent, had been among those who had watched Vincent Lunardi ascend from Moorgate. July 19 1821 had been chosen for this new balloon flight and the science of aerostation had now advanced to the point where coal gas had become the lifting agent; this was, in fact, the first ascent to be made with its aid.

On this occasion, the balloonist was an Englishman, Charles Green, one of the most important pioneers of the lighter-than-air craft and the vessel that he proposed to fly over London now was a fairly large one with a capacity of roughly 16,000 cubic feet. In the light of the event it was to mark, the balloon was given a special name, *George IV Royal Coronation Balloon.* The ascent from Green Park was uneventful perhaps because, at this point in time, the aeronaut was comparatively inexperienced and wisely did not attempt anything beyond his abilities, although during the flight an altitude of roughly 10,000 feet was reached. Perhaps the real value of this pioneer flight using coal gas was that its use was proven, a great step forward as the hydrogen previously used had been difficult to manufacture in sufficient quantities and, as a result, the cost had been prohibitive.

We are now at a point which is particularly rich in associations with flying of all ages past: and even the streets have something to offer, for a short distance away to the north may be found Audley Street, the part beyond Grosvenor Square which runs into Oxford Street. This is the thoroughfare which saw the descent of the first successful parachute drop in England when the Frenchman, André Jacques Garnerin, dropped from a balloon here on September 21 1802. Not that this was his first experience (this had taken place several years earlier over Paris in 1797, followed by three more) but the manner of carrying out the drop was the same.

The method was for a balloon to be sent up with the passenger already standing in the basket but, since no method of packing the canopy had yet been devised, this was hung loosely from its apex to the side of the envelope at its maximum diameter. The method of release was equally crude for it called for the parachutist to sever the lines which connected the gas bag to the car so that the latter dropped away, supported by the canopy. Although this system was all very well in its way, the problem lay in the design of the latter for as yet it was seen only as a complete umbrella shape with no vent hole at the top, with the result that the spillage of the air from about the rim caused violent oscillations, often to the extreme distress of the parachutist.

On this occasion from 9,850 feet over North Audley Street they had been more severe than usual and Garnerin, scarcely missing roofs at St Pancras, was feelling very ill when he landed. Unmindful of this a large crowd of admirers had gathered near the Jew's Harp tavern and were just about to lift him shoulder high in triumph when he was violently sick to the surprise of the onlookers. Among these, however, was a man of fairly mature years who straight away began to ponder the problem of the swaying which had made Garnerin so ill. In the fullness of time he came up with a new shape of parachute which, in the enthusiasm of his adventure, he failed to test first with a weighted dummy; an omission that was, in a different part of London, to cost him dear 35 years later.

Little more than half a mile from this historic spot is Hamilton Place and at No 4 is the headquarters of the Royal Aeronautical Society with its impressive and comprehensive library for serious students of man's conquest of the air. At 19 Park Lane stood Londondcrry House, at one time the home of the Royal Aero Club, a body probably unique in having been informally conceived in the basket of a balloon.

In the opposite direction one comes to Hyde Park Corner and, however unlikely it may seem, this junction can boast the exhibition of an aeroplane for

Heinkel 162, Werk Nr 120086, *on display in Hyde Park in 1945* (ATP).

it was here that in September 1950 a Lancaster I bomber was placed on display. The machine in question had been part of the strength of No 214 Squadron of the RAF then based at Upwood, Huntingdonshire, although there were detachments to various parts of the East which probably explain the white upper surfaces of the example chosen, *PA386:QN-Y*. The opportunity for the general public to examine aeroplanes in this manner was of long standing and the appearance of the Lancaster was just another example of the custom of displaying military machines in London parks. This had included such examples as the Friedrichshafen bomber, *1429/18*, which attracted crowds in St James's Park after the end of the First World War.

The end of the next conflict brought a much larger array of enemy machines to the area in 1945 when a small collection of aircraft had been made to form an exhibition opened on September 16. The few which were shown in Hyde Park brought some sharp criticism at the time, both from the press as well as the technical journals, in view of the very large numbers of Nazi machines that were to be seen flying up and down the British Isles. Nevertheless, the line-up of Nazi machines makes interesting reading and a surprising number of the machines on show still survive, as follows: Messerschmitt 110G-2/R3, Messerschmitt 163B-1, Werk Nr *191454*, together with a Messerschmitt 108B Taifun, a type that had temporarily been known as the Aldon during the war years to prevent confusion with the British Typhoon fighter. Other Nazi warplanes set out for examination were a Heinkel 162 *(120086)* a Junkers 88G-7, a Focke-Wulf 190A, *550214*, with a Fieseler Fi 165C to complete the powered aircraft, plus one sailplane, the bright yellow Grunau 2B advanced trainer.

These were not the only attractions offered to aviation enthusiasts for the same time a small show was also to be seen in Trafalgar Square, consisting of

another Heinkel 162, *Volksjager* (Peoples' fighter), a Fieseler Fi 103 flying bomb and an A-4, or V2 rocket. The nose of this latter dominated many of the photographs of the fly-past presented to mark the fifth anniversary of the Battle of Britain, a spectacle that was followed on the following day, a Sunday, by the display of aerobatics presented by Geoffrey de Havilland flying a Vampire fighter, the new and exciting jet fighter of the day. The best spot from which to enjoy this was undoubtedly Hyde Park and, attracted by the static display, large crowds had assembled there. It is an odd thought to us to ponder that perhaps some of those who were drawn to the Park that day had seen it in some of its earlier aviation moods, albeit unscheduled ones such as that, not 10 years before, when a Bristol Bulldog had made a forced landing there, an event also emulated by another that had come down in Southwark Park.

The following year, 1946, was that of the Victory Parades and to coincide with the one in London, another and more comprehensive display was organised in one of the Royal Parks, this time Green Park of balloon fame. This consisted of no less than 22 aircraft, all British in origin which could be viewed by the public from early June until Sunday the 16th. Perhaps because of its association in the mind of the public with the summer battle of only six years before, the most popular exhibit was a Hurricane fighter, despite it being a Mark IIc variety with its 20 mm weapons removed so that it resembled the earlier eight-gun model. The identity was *LF743*. With this was a Wellington TX trainer, *LP679,* converted from an earlier bomber role, and *RD773*, a Bristol Beaufighter X Torpedo aircraft. Two Airspeed products were the Oxford II trainer, *N6424*, and with it was a Horsa glider, *RN817*. Another machine which attracted much attention was the example of a Westland Lysander, a Mark III, *T1671*. This is because, at that time, little was known of the type's war-time activities as saboteur dropper so that its sombre finish, external ladder and long-range fuel tank were all certain to excite interest.

Striking an unexpected note by virtue of its all-aluminium dope finish amongst the darker camouflage hues was *LS326*, a Blackburn-built Fairey Swordfish and this was accompanied by an early Firefly IV, *TW688,* and the second prototype Hawker Sea Fury, *SR666*. There were other representatives of the Royal Navy: a Seafire Mark 47, *LA547*, equipped for F24 cameras and giving the public one of its first sights of the then-new contra-rotating propellers; a Blackburn Firebrand IV, *EK746*, with the finish that had not long been introduced, namely high-gloss camouflage, and Air-Sea Rescue was represented by *RD919*, a Supermarine Sea Otter biplane. Described as 'one of the very few remaining examples in serviceable condition' by the reporters of the period, there was also a Gloster Gladiator in the shape of *L8032*, which is still airworthy today, but with a different serial prefix, in the Shuttleworth Collection.

Naturally there had, even in that day to be both a Spitfire and a Lancaster bomber and in the former took the form of a Photo Reconnaissance version, *PS909*, with a pressurised cockpit and a Griffon 66 motor, with a Mark III version of the bomber, *ND677/G*, the suffix being the indication of the day that

Above left *Percival Proctor,* DX231, *following a forced landing in Hyde Park in 1946* (Imperial War Museum).
Left *The much-travelled Gloster Gladiator,* L8032, *as it appears in the care of the Shuttleworth Collection, Bedford, today.*

a guard was required to be put on the machine thus marked. Nor were light aircraft forgotten for there was the almost inevitable Tiger Moth, this being *T7307* from No 10 EFTS marked 'FDRK' and with it an Auster V, *MT363*. One more bomber posed its bulk against the background of grass and trees, this being a Halifax, converted for cargo use, *PP227*, officially designated a C Mark VIII. The size of the machines such as this dwarfed the remaining fighters but, despite this, considerable attention was shown in the Hawker Tempest V, *NV758*, and de Havilland Vampire *TG/277*, which, being a jet, would be voted star of the show along with the larger Gloster Meteor III, *EE235*, a representative of the mode of propulsion which had newly caught the imagination of many.

Much secrecy surrounded the identity of the pilot of the 'lone Hurricane' which led the Victory fly-past over London on V-Day itself. (History has since told us that it was Douglas Bader.) However there were very few enthusiasts at the time who did not realise that the machine was of the Mark IIc variety. Next came a trio of Sunderland flying boats from Calshot, found by No 201 Squadron and these were followed by three more 'heavies' this time Halifax Transports, actually converted bombers, Mark VIIIs from No 297 Squadron. Following these came a dozen Lancasters from No 35 Squadron at Graveley and even at the time they were earmarked to later fly to the United States to participate in the similar celebrations there; an occasion followed by a goodwill tour when the white upper parts provided a tempting background for hundreds of friendly messages, many inscribed in lipstick!

The next section of the Victory flypast over London was formed by 10 squadrons of various Marks of Mosquito, there being a grand total of 72 machines drawn from Squadrons 25/29, 219/264, 21, 107, 85/151, 811 and 248. Beaufighters from Thorney Island (254 Squadron) followed with Ford's 813 Squadron flying Firebrands, with 816's Fireflys from Lee-on-Solent. The next group of Seafires and Spitfires was led by the former, drawn from 807 with 587/691, 164, 165, 65, 287/567, 130, 91 and 41 following to make a total of 10 squadrons of various Spitfire types. An especially enthusiastic word was reserved for the de Havilland Hornets of 64 Squadron, the radio commentator describing the Horsham St Faith machines as 'the fastest propeller-driven aircraft machines' in use. Ahead of the jets came 54 Squadron's Tempest IIs with Tempest Vs of 3, 16 and 33 Squadrons. The rearguard was formed by some machines which seemed to herald the new world to come. They arrived with a roar—56, 74, 245, 234, 222, and 263 Squadrons' Meteors, some of them still Mark IIIs, to be followed by the Vampires of 247 from West Malling.

It is only a very short distance from our last point of call to the shopping centre of Oxford Street. It was here that, what almost amounted to a tradition had grown up, of the famous Selfridge's store exhibiting aeroplanes. These were from as early as 1909 when the Blériot XI, which had that summer crossed the Channel for the first time from France, was carefully crated after a brief and spontaneous exhibition at Dover in aid of police charities. Its new destination was to be the Oxford Street store.

All minds were set towards the future when the next important aeroplane was exhibited here, this time the Sopwith Atlantic, the wreck of which was placed on show for visitors to the roof of the building. This was the aircraft which was intended to carry pilot, Harry Hawker, and navigator, Mackenzie Grieve, across the Atlantic Ocean in 1919 from St John's, Newfoundland. From the

A Souvenir of the Exhibition in Selfridge's Roof Garden of the Sopwith Aeroplane (picked up in mid-Atlantic) flown by Messrs. Hawker and Grieve on their gallant attempt to cross the Atlantic, May 18th to 19th, 1919

Souvenir postcard sold to visitors who viewed the wreckage of the Sopwith Atlantic displayed on the roof of Selfridges in June 1919.

outset atrocious weather and an overheating motor doomed the project, despite several hours spent while the pilot attempted to cosset the failing power unit. Vanquished at last, they had been forced down beside the Danish vessel *Mary* which had, after 1½ hours, succeeded in picking them up from the dinghy that formed part of the fuselage. Not only was the *Mary* too small to attempt the salvage of the aeroplane, she also had no radio so that the flyers were given up for lost until the ship approached sufficiently close to the Butt of Lewis for a semaphore signal to be sent announcing the rescue of the two flyers.

Meanwhile the aeroplane was discovered, 10 days later, still afloat with the aid of the thousands of ping-pong balls with which it was stuffed to aid buoyancy. The American SS *Lake Charlotteville* hauled it aboard and later deposited it at London Docks from whence it was removed to the Oxford Street roof. It was exhibited here for the whole of the week beginning June 2, exactly a fortnight after it had taken off from Newfoundland and 12 days before the first non-stop crossing of the Atlantic was to be successfully completed.

It was June once more, but now the year was 1945 when another major show of an aviation nature was presented in a different part of Oxford Street. The venue on this occasion was the site of the John Lewis store that had been destroyed in the war-time bombing. On an excessively wet Thursday, June 21, an exhibition was opened here with the title 'Britain's Aircraft'. Entirely free, its intention was to show the achievement of the British aircraft industry during the preceeding six years. The visitors who made their way in via the entrance at the junction of Holles and Oxford Streets were in for a galaxy of surprises, not all of them due to the wealth of aircraft on display. At 1 pm, after a welcoming fly-past had loomed out of the grey clouds, minus the announced jet-fighters ('they can't get the paraffin', explained a man in the crowd), the workmen were still applying final coats of paint and attending to last minute details.

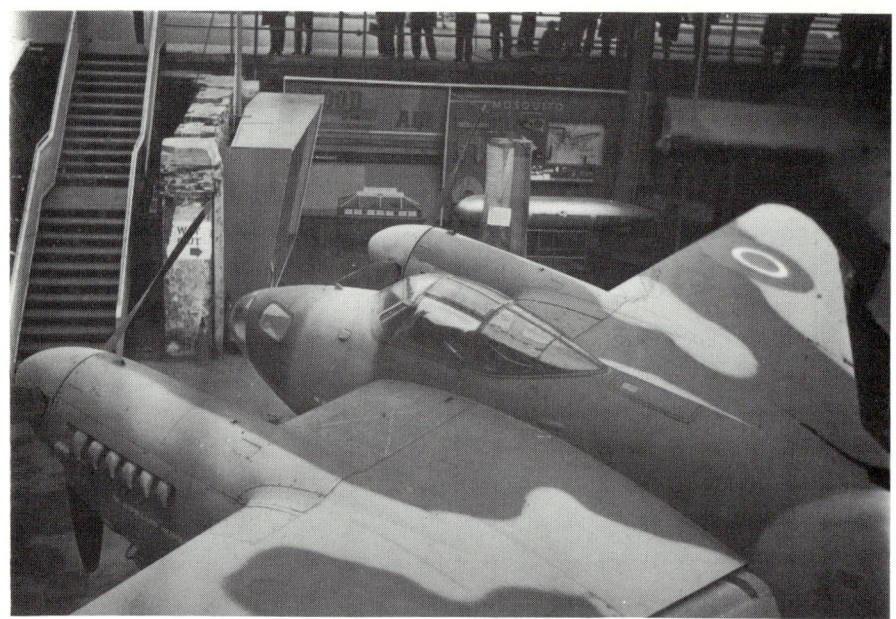

Above *The Mosquito, MM137, shown on the Oxford Street site of John Lewis* (ATP).
Below *Supermarine Walrus and Gloster E28/39 on show at the 'Britain's Aircraft' exhibition, 1945* (ATP).
Above right *The Gloster E28/39 during the preparations for the 1945 Oxford Street exhibition* (ATP).

Ten full-sized warplanes were crammed onto the site, the largest of them, shown intact, Handley Page Halifax III, *LV907*, a veteran of 128 raids, both by day and night, as a score on the nose declared. The show was organised in 29 sections dealing with as many aspects of the aviation industry as possible, mainly from the military point of view. Among these, many visitors would probably have voted the most interesting the final part which displayed, for the first time, the Gloster E28/39 jet aircraft, *W4041/G*, which now hangs in the Science Museum at South Kensington.

There was, at the time, a sincere interest in the new form of aeroplane power and this even spread to questions being sent to the popular radio 'Brains Trust' series, one asking 'Can jet planes fly in formation?' but on this occasion some little stir was caused by the erroneous placard which described the machine as 'the first jet-propelled aircraft in the World to fly successfully'. Undoubtedly the next most attractive feature for many was the display of a whole Supermarine Walrus, one of the comparatively rare Saro-built examples of Mark II variety which differed from the earlier model in having a wooden hull. Added to this fact was the presentation of *HD922* in a flooded area, although in fact it was not really floating.

In addition to the whole machines on view there were several parts and sectioned components, indeed the visitor was greeted on entry by the sight of a partially uncovered wing from a Wellington bomber in a superb silver-blue finish and, further in, the Lancaster was shown stripped of its skin with all the parts and services labled. This was a Mark III model which could be fitted with either the Packard-built Rolls Royce 28, 38 or 224 power plants.

However, although the public was treated for the first time to the sight of a 12,000 lb bomb or a Supermarine Spitfire LF IXE, *P1403*, displayed bereft of its port wing so that the upper surfaces could be seen as the machine was mounted on its side, it was the complete machines which commanded most of

Post-war civil aircraft were shown in the 1945 display only in model form. Here the rather austere internal layout of a Viking airliner is seen.

the attention. Among these was a de Havilland Mosquito, *MM137*, displayed in such a manner that it was possible to place a 4,000 lb bomb under the open bomb bay and, in addition, such relatively uncommon objects (to the general public that is) as streamlined wing-mounted bomb carriers were to be seen.

The representation of the army and the air was covered by two sections, one showing an Auster IV, *MT243*, fitted with a Lycoming 0-290-3 motor and standing on a section of BRC wire mesh runway. The other half of the army participation was supplied by a Horsa II glider of which the individual identity was a mystery because it was shown without the rear fuselage and also its outer wing sections. The main feature which distinguished this model from the Mark I version was that, through the opening nose section, it was possible to see stowed inside a six-pounder anti-tank gun. Other aircraft on the John Lewis site included a Bristol Beaufighter TFX, *LZ128*, fitted with both 25 lb armour piercing and 60 lb high explosive rockets to permit comparison of the two types of head, a Fairey Firefly *(MB627)* and *SN112*, a Hawker Tempest V.

Although this collection of military hardware was the only way in which the achievement of the years recently passed could be shown, the organisers had not forgotten to look ahead. To do so 10 models were chosen for display in a showcase representing the Avro XIX, Tudor and York, the Bristol Freighter and mighty Brabazon, the DH Dove, Handley Page Hermes, Miles Marathon, and two direct descendants of military machines, the Short Shetland and the Vickers Viking.

Chapter 5

Vauxhall to Hammersmith

In our day the name of Vauxhall conjours up nothing more than that of a district or perhaps a railway station, yet it meant something very different in times past, when this was that name given to a popular riverside resort. Vauxhall Gardens were opened as early as 1661 on a site at the back of what is now the Albert Embankment, near to the present Vauxhall Bridge, and in the following century they had become highly fashionable as a place for the gallants and where music might be enjoyed. One hundred years later the Gardens were past their best but were still looked upon by Londoners as a place of popular entertainment but, in order to maintain this, it was necessary to find some new delight with which to titillate the visiting public. The fact that the new sport of ballooning was increasing in popularity as a spectacle for the crowds provided the complete answer.

Mention has already been made in an earlier chapter of Jacques Garnerin in his capacity of pioneer parachutist but he was of necessity also a balloonist and, a little before his ordeal over North Audley Street, the crowds had seen him at Vauxhall Gardens. What they were witnessing on August 3 1802 was an ascent by the French man in the company of his wife and a Mr Glassford. An ascent of three persons simultaneously in a single car was enough to attract large crowds if only because it was a guarantee of a large and spectacular envelope, and there would have been those in the audience who would have known that Jeanne Geneviéve Garnerin shared her husband's interest in parachuting and was, indeed, the first woman to make a descent, on November 10 1798.

On this occasion, the lady was not to make a jump, but there was to be another surprise in store, for after the crowds had dispersed from the Gardens and gone to their homes, there was to filter back next day the astonishing intelligence that the aeronauts had remained aloft all night before finally landing at Frognal Place near Hampstead. All this was no more than a memory by the time of the next major ascent of a balloon from Vauxhall for it was not to take place until 34 years later, during which time ballooning had lost a little of its appeal as a spectator sport, but on this occasion the promise of a 'monster balloon' had ensured a good gathering. That it was large there is no question for, on this seventh day of November 1836, three persons had to be accommodated in the basket, the famous balloonist, Charles Green, who had gone up from Green Park in 1821 as part of that year's coronation celebrations, Thomas Monck Mason and Robert Hall, the Member of Parliament who was financing the project. This trip was intended to test a new type of trail rope.

A balloon about to rise from Vauxhall Gardens at the beginning of the 19th century (Pamlin Prints).

It was just noon when the gasp from the crowd announced that the balloon was away and the new vessel, completed that year from 2,000 yards of silk, made a brave sight against the November sky with its crimson and white panels and impressive name; the *Royal Vauxhall Balloon*. This spectacle was added to by the flags hung from the red car with its royal blue draping and the gilt bird's heads carved at the extremities. Not, however, that this was the first use of the coal-gas-filled *Royal Vauxhall*. It had been used on three previous occasions, the first of these being witnessed by Lord Palmerston when nine persons, including Mrs Green, had been bourne aloft, as already related.

By now, it had become customary to make more serious provision for voyages of this sort so that, secure in the knowledge that an ample supply of food and drink, plus warm clothing, instruments, and signal lights were aboard, the three felt no apprehension when the prevailing wind took them steadily southwards. After some slight altitude adjustment they crossed the English coast at Dover and, not long after, were above Calais, but by now the short winter day was beginning to fade and, by the time they reckoned to be over Belgium, night had fallen. The feelings that all experienced at this point were somewhat peculiar for, despite the sight of the illuminations below and over Liège, the glow from the blast furnaces, there was a sensation of being conducted through a solid substance which one likened to 'black marble' that seemed to soften to admit them into its cold embrace.

After several hours aloft the men had lost all idea of their whereabouts and could only guess at this from the occasional clear sights of the terrain below, part of which was covered in snow, so that 17 hours after leaving Vauxhall it was decided to land, by which time they had crossed the Rhine north of

Koblenz. Landfall was made without mishap near to a forest in the early hours of November 8 and in a few minutes they were approached by a group of farmers who informed them that the field they had chosen was in fact only eight miles from Weilberg in the Duchy of Nassau, so that the three were some 480 miles from the point of departure.

The longest balloon voyage on record, and one that was to remain unbroken until 1907, ended with celebrations in the town while the balloon, renamed the *Great Balloon of Nassau*, was later transported to Paris for exhibition. It was announced in 1837 that Charles Green was to attempt another ascent from Vauxhall Gardens and, on this occasion, a parachutist was to make a drop from the vessel. The name of this pioneer was Cocking, the same Robert Cocking, a water-colour painter who had watched Garnerin in September 1802, and witnessed the distress that the swaying canopy had caused the experimenter. Now a man of 61 years of age, Cocking had spent many years of amateur experiment in attempting to solve the problem, but always he had failed to use a full-size parachute by means of which tests with a dummy might be carried out.

The basis of these investigations was the belief that ideally the shape of a canopy should be that of a cone, used inverted with the person suspended underneath, near to the downward-facing point, the actual fabric being supported with wooden ribs. It was such a contraption that Green was persuaded to take aloft—much against his will, to allow Cocking to try out his theories personally, although the stipulation was made that Cocking and not Green, be responsible for the actual release. On July 24 1837 Vauxhall Gardens was a scene of bustling activity, but it was early evening before conditions were regarded as being just right. The balloon was again the *Great Nassau* and, in addition to the weight of Green and Cocking, another occupied the basket, Edward Spencer.

The time was exactly 23 minutes to eight when the ascent was made with Cocking suspended under the balloon's car some considerable distance from it. As a precaution (in case the weather deteriorated to the point where a drop was impossible) a special tackle was rigged whereby the parachutist could haul himself up to the car if this was needed. Another safety measure had been the tests with the balloon loaded with 650 lb of ballast to determine its lifting potential.

Great Nassau was over Blackheath at a height of 5,000 feet by seven minutes to eight, but it was only with difficulty that this altitude had been reached after the discharge of 50 lb of extra ballast. At this point Green leant over the side and asked Cocking if he was quite comfortable and if the trials so far seemed to bear out his expectations. The artist's reply drifted up from under the basket.

'Yes! I have never felt more comfortable or more delighted in my life.' There was a moment of silence before the speaker continued. 'Well, now, I think I shall leave you.' At these words, Charles Green again bent over the side with his megaphone to his lips. 'I wish you a good-night and safe descent if you are determined to make it, and not use the tackle.'

Almost at once a jerk was felt by the two men in the car which announced that Cocking had pulled the release: a slight pause followed during which it was evident that the parachute had failed to disengage, so that there was now a second and more determined tug. As the parachute dropped cleanly away, the balloon, suddenly bereft of the weight below, was shooting upwards 'with the velocity of a sky-rocket' so that it called for all Green's skill to regain control and it was only his presence of mind that allowed him to do this or else both he and Spencer would have been taken to their deaths.

Meanwhile, observers on the ground noted that the parachute was falling rapidly with no other sign than a slight upward bend of the canopy and all was going well when it was lost sight of in a cloud. It was still behaving normally when it re-appeared, but, about 40 seconds later, the ribs gave way and the rim of the canopy collapsed so that, oscillating wildly under the broken cone, the form of Robert Cocking was seen to be falling at a frightful rate to plunge finally to earth in a field in Kent near to Lee Green. He was still alive when the first assistance arrived but died after only a few seconds. The coroner's report had this to say: 'We find the deceased, Robert Cocking, came to his death casually, and by misfortune, in consequence of serious injuries which he received from a fall in a parachute of his own invention and contrivance which was appended to a balloon, and we further find that the parachute, as moving towards his death, is deodand and forfeit to our Sovereign Lady the Queen.'

The site of the historic Vauxhall Gardens where, as early as 1784, King George II had watched Vincent Lunardi make an ascent in a balloon, provides the limit of aviation interest in this area. However our study of south London would remain incomplete if we did not spare a glance in the direction of Tilbury. A journey would not be called for. (This is why no more than a mention of a neighbouring district needs to be made.) This area was the subject of an interesting proposal that was announced at the beginning of 1938 concerning a projected Empire Airport which, it was agreed, should be situated as close as possible to the centre of London. This suggestion makes the present day search for sites to augment London Airport look strange.

The proposed Airport was to make use of the River between Gravesend and Canvey Island on that part known as Lower Hope Reach, one of the chief attractions being that, adjoining the Thames on the Essex side, lay some 300 acres of ground. In addition, close to Mucking village was the Tilbury railway line which could easily be extended, as could the Muckingford road, to provide a link with that from London to Southend.

At the time of the issue of the proposals, the open land was in the process of being made up with the capital's rubbish, and reports were at pains to point out that the area offered facilities for both land and maritime aircraft while the approach was singularly free of obstructions, even the nearest high ground being sufficiently distant as to present no problem. With regard to the river facilities it was considered that the reach at Hope point was almost ideal since the prevailing south-westerly wind blew down the four miles of the longest run at a sheltered spot and even the shortest run of a single mile at low tide was capable of extension with a programme of dredging, so that there were natural facilities for take-offs in every direction.

It was anticipated that terminal buildings could be erected on the western shore opposite Hope Point. This would be served, along with the hangars, by a new road and rail connection north of Tilbury. Adjacent to the administrative buildings it was planned to construct a new dock, partially under cover. It was expected that flying boats could taxi, at high tide and under their own power, into one of the four bays for maintenance and overhaul. Alas, all these heady dreams came to nought and were forgotten as war overtook Europe.

Following an indirect route towards the western part of London brings us into Eaton Square which, apart from being the area where the RFC Hospital could

Eaton Square contained one of the London Royal Flying Corps' hospitals.

be found, was also the home of Lord Edward ('Ned') Grosvenor, the larger-than-life son of the first Duke of Westminster. His love of adventure had sent him straight from public school to the French Foreign Legion but, when this failed to satisfy, he entered the Royal Horse Guards, thus making his early career no doubt unique. His connection with aviation is a strong one for it was the aristocratic 'Ned' who founded No 601 (County of London) Squadron of the Auxiliary Air Force in 1924 and it is said that an unexpected connection between the flying world and White's was formed when the first members of the new unit were recruited from the membership of that illustrious club. In so doing, Grosvenor was in effect re-establishing his connections with aviation for, during the years of the First World War he had been an RNAS Flight Commander at a time when his London home was still at 2 South Street, Park Lane.

One of his peculiarities in this duty was his habit of giving directions to new pilots, not by a compass bearing from the base at Eastchurch on the Thames Estuary but by *pointing* in the right direction and giving instructions as one would on the road, namely when to turn right and left. History tells of the classic occasion when the index finger, denoting the direction to Belgium, seems to have been misunderstood for the unfortunate pilot was to finally make an emergency landing . . . in Hammersmith Broadway where his aircraft caused no end of interruptions to the tram service! However, none of this is to say that the character of 'Ned' Grosvenor was, although at times perhaps unconventional, anything other than that of a charming and robust enthusiast. He led his men as much by his own open and honest example as by recourse to service discipline.

Although the town headquarters of this auxiliary unit was at 54 Kensington

Park Road, W11, the airfield was at Hendon for the greater part of the squadron's pre-war existence, to which base it changed in January 1927 after 14 months at Northolt. These formative days had been spent at a light bomber unit, for which role its equipment had at first been DH9As, to be exchanged for the Westland Wapiti towards the end of 1929. This same year saw the sudden death of the inimitable Lord 'Ned', at the age of 37, after a short illness, so that the new Commanding Officer was Sir Philip Sassoon, another resident of Park Lane; all entirely in keeping with the title by which No 601 was unofficially known, namely 'The Millionaires' Squadron, there reportedly being six members in this fortunate position at one time.

The last type to form the squadron's equipment before it changed its role was the Hawker Hart but, by 1934, these were to go and in their place appeared Hawker Demons. These were not taken on strength, however, until two years after the unit was designated a fighter squadron so that for a period of 24 months it was somewhat unusual in being a fighter squadron exclusively flying light bombers! With the coming of the Second World War, Gauntlets, Blenheims, Hurricanes and Spitfires formed the successive equipment with which it went far afield to Malta and Italy as well as participating in the Sicilian campaign.

When the fighting was over 601 still retained its Spitfires, now LF16Es until jets arrived in the form of the de Havilland Vampire at the end of 1949 and these were later exchanged for Gloster Meteors. Sadly this was to be the last fighter that the squadron was to fly. In a wave of Government cuts in expenditure that axed the entire Royal Auxiliary Air Force, and on a chill but clear day at the beginning of March, 1957, the squadron marched past for the final parade. Prince Philip took the salute outside the Kensington Park Road headquarters, while in the watery afternoon sunshine the flag of 601 flew, appropriately, at half mast.

Returning our attention to Belgravia it is only a very short distance to the next port of aeronautical call which may be found in Knightsbridge (at the top of Brompton Road) in that advertised 'different world' of Harrods. The first main connection that this excellent store enjoyed was an indirect one, and perhaps happily so. It concerned the alarming number of casualties that were being suffered by the Royal Flying Corps in France so that the outspoken Member of Parliament, Noel Pemberton Billing, was moved to declare in the House of Commons that airmen had 'been rather murdered than killed'. The cause of this outburst was the number of aeroplanes of the officially designed BE2 type and its current variants that were in use, despite their unsuitability for the majority of fighting roles due to their 'inherent stability'. Naturally such a declaration as this would have caused a degree of public unrest at any time but in war it was particularly grave. The Government therefore set up a Judicial Committee to investigate the charges and it was this body that even now is sometimes referred to by historians as the Burbidge Committee from the name of its Chairman who was also head of the Board of Harrods at the time.

A happier association enjoyed by this business house was to take place immediately after the end of the war, at just about the time that a new type of motor vehicle was appearing on the roads of London. To deal with this topic first, the lorries that were currently attracting the attention of small boys and their fathers were at first only six in number but they seemed to be everywhere. These were part of a purchase, in 1919, of surplus RAF Leylands which had seen service in the Middle East by Sainsbury's and were the first of that company's motorised fleet.

The main attraction of such vehicles was that they were extremely robust and relatively cheap but although nothing could be done about their solid-tyred road wheels, Sainsbury's engineers were of the opinion that some improvements were possible. Strangely, no one thought of improving the lot of the drivers who were condemned to sit under a canvas hood that, although it certainly kept the worst of the weather off, was a long way from giving an atmosphere of snugness in the winter. It was the lighting that occupied the attention of the engineers, and the headlamps, formerly oil-burning, were replaced by electric ones. Perhaps this was a wiser move than it at first sight appears for these pulsating vehicles (which had a reputation for excessive vibration) were to be seen rumbling out of the Blackfriars yard at between 2 and 3 am in the morning, bound with supplies for distant branches.

That same year of 1919 the Blackburn Company produced a most interesting little civil aeroplane with an eye to catching the anticipated sports flying market. It was this machine which formed the focal point of a display in Harrods, its mechanical attractions being augmented at the time when a number of photographs were taken of a number of girls chosen for their looks and charm. They were grouped round it, clad in the flying kit, or at least an idealised version, of the day.

Power for the original design was supplied by a 40 hp ABC Gnat motor and it was mounted in front of an unusual seating arrangement for the time, namely the positioning of the two occupants side by side. At a price of £450 the Blackburn Sidecar monoplane, a name seemingly derived from its layout, was an expensive item in an age when few men earned £5 per week, even in a professional field. It is little surprise therefore that only this one example was built and the buyer decided it was underpowered and so had it fitted with a 100 hp Anzani engine not long after purchase.

Before leaving this immediate neighbourhood there is one other, half-forgotten connection that should claim our attention. It may be found not far from this spot at Imperial Court, Basil Street, the thoroughfare that connects Sloane Street with Hans Road. Here was the Vickers drawing office at an address it was to occupy into the early part of the 1920s.

Take either of the main roads from this area, and a short journey towards the west will bring one to Exhibition Road and inevitably this puts one in mind of the National Aeronautical Collection in the Science Museum, the oldest one of its type. The first contribution was made by Sir Hiram Maxim in 1896 when he gave a steam power unit, a massive airscrew of 18 feet in diameter and a model of his flying machine. Although many of the aeroplanes on view are much older, the main nucleus of the exhibition was gathered during the 1920s. At that time the machines were housed in the ground floor gallery beyond the one which is approached from the main entrance in Exhibition Road. It remained here, apart from a fairly brief sojourn in the old Empire Institute building, until it was housed in the present gallery that was opened in 1963.

During the days before the collection was dispersed to the west country, as a safeguard against war-time bombing, there were not only some of the aircraft currently on show but some others that have since been lost, including the Bristol Bulldog which was later rebuilt to fly and crashed as *K2227* at Farnborough in September 1964 in the hands of Mr J.I. Williamson. During the days of its display at South Kensington this IIA version of the famous fighter proved to be the former demonstrator, *G-ABBB*, which was all silver with a blue decking and was displayed partially stripped to show the internal structure.

Another vanished exhibit was the genuine Wright Flyer which had been on show for 20 years from 1928 and the present replica was made when the return of the original was requested but such old stagers as the Cody Biplane, acquired in 1913, and the Roe Triplane, which arrived in 1925, remain. Another display of a temporary nature was the Supermarine S6 seaplane, *N247*, which won the Schneider Trophy for Great Britain at 328.63 mph in September 1929—not to be confused with the 1931 outright winner, the S6B, *S1595*, on show today. The earlier seaplane was exhibited immediately after its triumph on a roped-off dais constructed over the main steps inside the entrance.

It is clearly impossible to give detailed histories of all the machines in the care of the Science Museum but an exception must be made for one since it is the Hawker Hurricane, already described in Chapter II, when it was almost a regular on Horse Guards Parade, the example in question being *L1592*. The first recollection of this machine which ex-members of 615 (County of Surrey) Squadron still retain is that it probably arrived at Kenley in June 1940 from No 56 Squadron in which it had previously served. By now it was regarded as being something of a museum piece since the fabric-covered wings that it still retained had long been replaced by metal-skinned ones.

However, when the squadron was scrambled on Sunday, August 18, Pilot Officer David Looker sprinted for this reserve machine more usually flown by Sergeant Pilot Haughton. It was, like the young Pilot Officer, a veteran of the Battle of France but any machine was better than none and his usual Hurricane was unserviceable at the time. Once in the air with the rest of 'B' Flight it looked as if the enemy was not to be found when, quite suddenly, the plane on Looker's right went down in flames, after a well-aimed burst from an undiscovered Messerschmitt 109. The unseen attacker now turned his attention to *L1592* and a burst of fire aimed at the tail quickly put the Hurricane into a dive from 25,000 feet. From this it seemed reluctant to recover and the pilot rejected his first idea to bale out and decided to do his best to save the precious fighter. To limp back to Kenley was out of the question but Croydon, its satellite, seemed to offer an alternative landing ground so the pilot thereupon turned to make an approach over Purley Way from the east. As the green of the airfield grew larger he selected flaps and undercarriage down and was relieved to find that the hydraulics were still in working order.

The ground gunners at Croydon had just received a fierce straffing and they were consequently distinctly trigger-happy. Unfortunately aircraft recognition as taught to soldiers still left something to be desired. When an ugly, sinister, angular shape appeared, approaching low and with determination, it could mean, they reasoned, only one thing—a marauding Stuka dive-bomber. No one paused for a second look and immediately the guns sent up a curtain of hell against the seeming raider. Astonished at the reception, Looker had no alternative but to hold his machine, already damaged by the enemy, firmly on course and, as much by luck as anything else, the curtain of fire left the Hurricane undamaged. It was with a feeling of relief that the pilot heard the welcome rumble of the wheels over the grass, but his pleasure was short lived, the onslaught by the gunners had taken its effect on him and hardly had the fighter travelled a few yards when its career was abruptly stopped as it tipped up

In the foreground the Supermarine S6B racer and the Gloster E28/39 Sea Hawk frame Hawker Hurricane, L1592, restored to 615 Squadron markings in the Science Museum (Crown Copyright).

on its nose and from the cockpit, at an ignominious angle, David Looker had to slide to the ground. He was then taken to the local Croydon General Hospital where he was attended by Dr John Wainright who was later to remark that in his view the machine was 'almost tail-less'.

Seven days later, *L1592* was transferred to No 13 Maintenance Unit but its subsequent career is obscure. It was variously described as being seen in a seldom-used hangar at Biggin Hill and it may be the same machine reported to be in the almost empty Science Museum during 1944. Whatever the train of events, 14 years after its last recorded use in battle, it was restored to the appearance it now possesses by Hawker's Langley factory. Here too is the

Above *SE5a,* F939, *in the Science Museum, South Kensington, was at one time a smoke writing aircraft.*
Below *The Supermarine S6B seaplane racer which won the Schneider Trophy outright for Great Britain in 1931.*
Above right *Amy Johnson's DH Moth 'Jason', named after her father's brand of kippers, on show in the Science Museum* (L.J. Dickson).

Gloster E28/39 that we mentioned last at the John Lewis site in Oxford Street and, of course, the Vickers Vimy. This had brought John Alcock and Arthur Whitten Brown across the Atlantic Ocean non-stop in 1919 to a London that mobbed them on their way to a luncheon at the Savoy on June 20. Afterwards they received from the hand of Winston Churchill the prize cheque, offered by the *Daily Mail*, for £10,000. Brown used to come annually on the anniversary of the flight in the evening of his days and stand bareheaded and silent in the shadow of the Vimy's wingless bulk, in the original gallery.

Fokker EIII monoplane, *210/15*, now suspended in skeletal form, fell victim to a punctured fuel tank and consequent petrol loss during the First World War and underwent a series of tests before being circulated as a battle trophy and eventual exhibition here. Appropriately it is not far from an SE5A fighter presented to the Museum by its last owner who used it to write in smoke across the sky such advertisements as 'Buick' and 'Persil' in the 1920s and '30s, so that its stay in the care of the Science Museum is shorter than that of the Avro 504K which came here only two years after it was taken on charge by the RAF in March 1918.

Perhaps it is the record breakers that are the most evocative sights under the mock hangar roof and among these one of the most compelling must be Amy Johnson's *Jason, G-AAAH*. It was in this second-hand de Havilland Gipsy Moth that the typist from Liverpool flew 10,000 miles alone from England to Australia in May 1930, and it was presented to the Museum the following year. True, its appearance today is not exactly the same as it was during the epic flight. At that time it had the front cockpit blanked off as it was flown from the rear seat, and a spare propeller lashed to the centre-section struts altered its outline somewhat and the old style AA badge on the cowling was yet to come.

In addition to the aircraft described, the collection includes other full-sized examples (all of which are in the Museum at South Kensington, plus two further small groups, one lot being in store and the other in collections elsewhere). They include a Lilienthal glider with a replica of the Pilcher Hawk aircraft and a genuine Antoinette Monoplane similar to that in which Hubert Latham unsuccessfully competed with Louis Blériot's attempt to be first across the Channel. Bearing a strong resemblance to the Blériot machine is the JAP/ Harding Monoplane and research work in the 1920s is represented by the Westland/Hill Pterodactyl IA with a Cierva C30A. In addition Second World War aircraft include a Spitfire IA, Focke-Achgelis Fa-330, Messerschmitt 163B-1, while the dawn of missiles is shown by a Fieseler Fi 103, V-1 flying bomb, together with the British precursor of VTOL, the Rolls-Royce 'Flying Bedstead' near to a Saro Skeeter 12. Apart from sundry items of space equipment and mock-up flight decks there is a wealth of aero-motors of all types, together with sundry airscrews from every age of powered flight, plus a very large collection of representative models, many of them the work of that master of the craft, the late V.J.G. Woodason.

One cannot continue far west along Kensington High Street without coming across the exhibition hall known as Olympia and perhaps wondering if it has any aviation past. That it has is no great surprise for the present-day Boat Shows and the former London Motor Shows are an indication that aeroplanes must have been shown as static exhibits at some time. The first of these took place in 1909 under the sponsorship of the Society of Motor Manufacturers and Traders and another took place two years later combined with the Boat Show. By 1914 there was to take place the fifth of these and, in fact, many were of the opinion that this was the high water mark of these shows. Even so, they were continued after the Armistice until 1929 when 94 exhibitors took stands, warranting a catalogue of 324 pages. This included such non-technical information as the programme of music to be played each afternoon and evening, with a daily change of repertoire by the String Band of the Royal Air Force under the baton of their Director of Music, Flight Lieutenant John Amers, MBE.

To the present day enthusiast the Royal Tournament is associated with Earls Court, although this has not always been the venue in days gone by. It is not too readily known that a rota ensures that the Services each take in turn the provision of a theme for the display, this ensuring a balance of emphasis over a period of time; 1976 was a case in point when the Grand Finale marked the birth of the Royal Air Force with songs and tunes of the time. Frequently replicas of aircraft, indistinguishable from the originals in the arena have been used while at other times the genuine article had been pressed into service. One such event was 1953, the year of the Queen's Coronation and on this occasion the much-travelled Sopwith Camel from the Imperial War Museum was on loan to the organisers. This continued its roving life, stretching as far back as January 1934, when it was shown at the White City in connection with the Schoolboys' Own Exhibition there and explained by the redoubtable W.E. Johns, creator of 'Biggles'. In similar vein was 1962 when, with the aid of a combination of mock-ups and the real thing, the story was depicted of the whole pageant of flight. The London connection was doubly pertinent since the first dress rehearsals were conducted in the south east, at Surrey's Kenley Aerodrome where a 1911 Vauxhall car gave an added air of authenticity when it glided past with its load of 'staff officers'.

The area in which we now find ourselves has a number of slight aeronautical connections in each direction, in Brompton Cemetary lies Reginald Warneford, VC, who was to die tragically a very short time after becoming a national hero when he brought down the *LZ37 en route* for a bombing attack on Berchem Ste Agathe, over Bruges in 1915; his resting place at Brompton is marked by a singularly ornate headstone. Likewise, at 235 Hammersmith Road, was the firm of William Cole & Sons Limited, the contractors. They were known as Cole Aircraft at much the same time and are therefore representative of the many hundreds of manufacturers who produced aircraft and parts not only in London but throughout the country. The lesser-known organisations such as this shared the market with those whose names were household words such as Hoopers of St James in another area.

However, our final focal point is a little way off, towards the river and behind King's Road, part of the site being that of Lots Road Power Station, the road itself being an extension of Cheyne Walk. At the intersection of these two is the first clue to the nature of the interest here for the way that joins the others is named Cremorne Road. This is a name that immediately conjoured up a picture of the successor to Ranelagh, as the claim went, that was established as a place of public amusement in 1830 when Baron de Baufain founded the National Sports Club in the grounds of Cremorne House that he had just purchased.

Between that time and the closure of the Gardens 47 years later there were hundreds of balloon ascents made from here and the first of these was made by none other than Charles Green who in 1837, using the *Great Nassau* once more, went up in the company of a lady and a leopard! This was the first of 526 ascents he was to make from Cremorne Gardens. Eight years later he took with him his wife and Lord George Beresford and the Drury Lane clown, Tom Mathews, plus two others, the latter gentleman in full costume and stage make-up. Two hours later, after reaching an altitude of about 6,500 feet, they landed on a soft but pungent spot; Tottenham Marsh, but little mishaps such as this were all part of the sport and did not deter Green from later making a most dangerous trip in an electrical storm from the same Gardens, but he only managed to reach Harrow.

In September 1852 Madam Pontevin was fined for taking a heifer aloft with herself seated on its back dressed as Europa, while 12 years later saw a revival (not for the last time) of hot air balloons. A Mongolfier of 95 feet diameter was demonstrated so that visitors had the chance of comparing this in their minds with the next spectacle the Gardens presented, namely a 2,000-foot long dirigible with propellers for steering. During this time children's balloons were a common sight in the gardens. At the other end of the life scale there is record of a lady who celebrated her 100th birthday by an ascent in the resident captive balloon which was wound up and down by means of a 200 hp steam engine. One hopes that the usual fee of 10 shillings (50p) was waived by the organisers on this occasion, it would have enhanced her enjoyment of the view she had from 200 foot altitude.

Not only balloon ascent but also parachute drops were to be seen over the Gardens and the most celebrated of these was that attempted by the Belgian, de Groof. The apparatus was in fact more closely allied to a hang-glider with a wing span of 37 feet. Suspended under a balloon on a 30-foot line he cast off, opened the wings and made a safe landing although out of sight of the crowds as the contraption of glider and balloon was driven to Essex where the final separation was made over Brandon. It must have seemed that the jinx of Tottenham Marshes, which had claimed the life of French parachutist, Henri Latour, when he dropped from W.H. Adams' balloon in 1854, had at last ended and perhaps in mind of this, de Groof made a new ascent on July 9 1874.

Although the balloon was reported to be a captive one on this occasion it seems to have broken free because we are told that it drifted towards St Luke's Church, Chelsea, at about 900 feet. Confident that the earlier success would be repeated, de Groof cut the rope by which he was suspended and attempted to control his descent with the steering tail some 18 or 20 feet wide. Alas, on this occasion the wings failed to properly unfurl and they finally collapsed so that the Belgian crashed in Sydney Street. Technically alive when he was picked up, de Groof died almost at once. The balloon drifted on to Springfield where it landed on the railway line, narrowly escaping an oncoming locomotive. Sightseers undoubtedly preferred to remember the happier occasions such as that on the night of August 24 1954 when a Lieutenant Gale, RN, let off fireworks from the car of his balloon, notwithstanding the quantity of explosive gas a few feet above his head!

Chapter 6

Hendon and its area

Time has accumulated a surprising number of aerodromes and flying fields to the north of the River Thames although only a percentage of these survive today and not all of them now fulfil their original function. Many were established as part of the London Air Defence System against Zeppelins and lasted little longer than their immediate requirement, and at least one is entirely fictional! Some were no more than Night Landing Grounds, part of a pattern of such fields scattered about the capital on both sides of the river, the nearest on the south side being Wimbledon Common.

It is not proposed to cast our net too far north of the Metropolis but even so it is worth mentioning casually that Radlett, established in 1929 and extended for use in the Second World War, was the home of the first SBAC shows in the immediate post 1945 years before the name of Farnborough became synonymous with the display. Like Radlett, a manufacturer's aerodrome, there lies not far away, Leavesden (Watford) dating from 1940 while the similar Stag Lane (Edgware) was associated with the de Havilland Aircraft Company and its predecessors from 1917 until its closure in 1932. Five years later (in 1937) and a few miles away, the private aerodrome at Elstree, otherwise known as Aldenham was established.

Almost due south east of this site and the other side of Hendon which we propose shortly to deal with at length, is supposed to have lain the aerodrome at Kilburn. Even residents of long standing will be perplexed by an enquiry as to the whereabouts of this field and additional information that it saw the maiden flight of the Central Aircraft Company's twin-motor civil transport in 1919 will elicit no response. The fact is, that although the Centaur was certainly built at the works of the old-established woodworking firm's Kilburn subsidiary, there never was a Kilburn Aerodrome and the initial tests of this seven-seater would have taken place at Cricklewood. This was another manufacturer's airfield not a mile distant, and one which had been established by the Handley Page concern, despite the built-up nature of the area even in 1916. Nearby were once the now largely-forgotten Wembley Park and Park Royal, both of which vanished in 1912, the former after only three years existence. Wormwood Scrubbs was for nine years devoted to airship construction and like Park Royal opened in 1910. About two miles to the west was Acton which was only involved in aircraft construction for about 12 months in 1915-16. Equally short-lived was Westpole Farm, from 1918 to 1919.

In the east, below Chingford First World War aerodrome that closed after a

six years use in 1921, lay a couple of landing grounds founded in 1909 and closed two years before the outbreak of the First World War, Lea Marshes and Barking Creek. Far more important and of greater historical interest were undoubtedly the twin aerodromes of Hainault Farm and Sutton's Farm at Hornchurch. These were established in an attempt to meet what was regarded as a terrible menace—the airship bomber. There is no doubt that this was seen as a major threat but it must be admitted that the attitude of the general public, namely that in the Zeppelin there existed a weapon against which there was no defence, had in no way been helped by the official approach. The Authorities were quite capable of producing posters announcing 'It is far better to face the bullets than to be killed at home by a bomb', which was nothing less than an admission of defeat. In practical terms there were real measures being taken, too, to allay public terror and one such step was the distribution about London of small units of pairs of night interceptors which were shortly amalgamated to become No 39 Squadron.

The night of Saturday, September 2 1916, had been chosen for a concentrated attack on London and 12 airships were to take part, drawn jointly from the Army and Navy. Among the former which approached from Belgium across the Thames Estuary was the *SL11* which had made its first flight as recently as August 2. The Naval airships meanwhile, made headway across the North Sea and crossed the coast in the region of Norwich. The first craft to approach London was Ernst Lehmann's *LZ98* and the first bombs were dropped at 1.15 pm on Gravesend which was mistaken for the London Docks.

Meanwhile several BE2Cs had been sent off. Lieutenant Ross, who flew the BE12 variant, left North Weald at about 11.10 am. Lieutenant Brandon's patrol was to take him to a height of almost 10,000 feet. Lieutenant Sowrey took off a few minutes after 1 am on the Sunday morning to be followed, some 15 minutes later, by Lieutenant Hunt who was to be rewarded by a sight of *SL16* a quarter of a mile away and below his own altitude, which he soon lost. Other pilots climbing into the night sky were Lieutenant Mackay who was to patrol in the vicinity of Joyce Green, the three-year-old aerodrome on Dartford Marsh, and Lieutenant Robinson, formerly an RFC observer, in France where he had been wounded. It was this officer who suddenly spotted the escaping *SL16* at just under 14,000 feet but, after a short chase, the airship was lost in the cloud.

Following the encounter with this aircraft the next was with the *SL11* and it is a curious coincidence that the captain, Wilhelm Schramm, had been born in England 20 years before to Karl and Maria Schramm at what is now 9 Victoria Way, SE7 although at that time it was known as Victoria Road and reckoned to be in Old Charlton, Kent, when Karl was acting as the London representative of the Siemens electrical firm; a post which he retained until a stroke meant his return to Germany where he died in 1900. Airship Shutte-Lanz 11 had made landfall at almost a quarter to 11 over Foulness Point and had then taken a wide sweeping course across the country which had embraced Great Chesterford, Royston and Luton before turning south east when the first bombs were dropped not far from St Albans at London Colney. These were followed by small numbers distributed over North Mimms, Littleheath, Cockfosters and

Left *A formation of Sopwith Camels over Lea Valley Road, E4. The former Chingford Aerodrome is now a reservoir* (Crown Copyright).

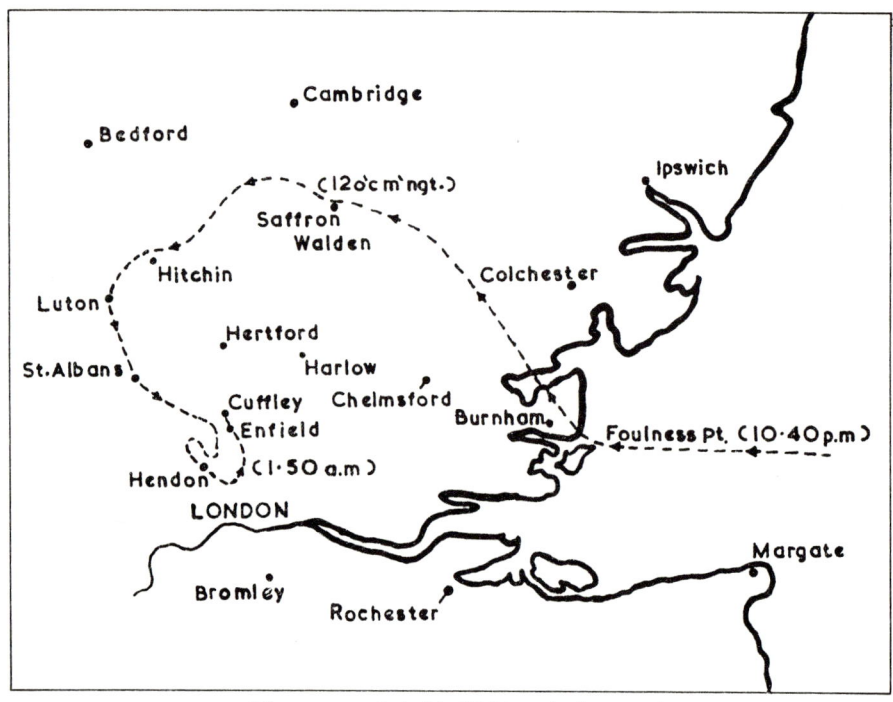

The course of airship SL11 *on its last sortie.*

Hadley Wood. While this was going on, the defences were on full alert and it was the probing searchlights which had given a heightened awareness of what was at the time called a 'raid feeling'. Among the half dozen searchlights which now traversed the northern sky were those based in the Parks at Finsbury and Victoria and it was this pair that suddenly picked up *SL11*. The resultant hail of fire causing 'Willy' Schramm (the name by which he liked to be known) to alter course towards the north so that bombs were dropped over Ponders End, Edmonton, Forty Hill and Turkey Street.

The other William (Leefe Robinson) was by now at a height of 10,000 feet after a 53 minute climb. Over Woolwich at about 1.10 am his attention was caught by a concentration of searchlights which had picked up a Zeppelin *(LZ98)* but were finding difficulty in holding it as cloud had gathered in this area although on the whole it was a beautifully clear night. In the hope of cutting off its retreat, Robinson climbed a further 2,000 feet, putting the airship about 800 feet below him, sacrificing speed in order to keep his height. He then gave chase, slowly overtaking his quarry but, after some 10 minutes, the Zeppelin evaded the lights and was lost in cloud. Fifteen minutes' search by Robinson, failed to find it again.

An hour now passed and the searching BE2C was north of Woolwich when the pilot noticed a red glow north east of London and he flew towards it when, from his altitude of 12,900 feet, he was rewarded by the sight of another airship caught in searchlights in the north east quarter and he made towards the shape, this time putting the nose down to sacrifice height in order to gain speed. As he did this he lifted (onto the Lewis gun through the aircraft's centre section) the

These picture postcards showing artists' impressions of the event were typical of the souvenirs sold to commemorate the first destruction of an airship over England.

first drum of mixed incendiary and explosive ammunition of a type that had only been introduced that summer.

About 800 feet below the gigantic cigar shape, Robinson distributed the whole of this first drum along its length from bow to stern from one side, but without apparent effect so that for the next attack, he remained aft and at a range of 500 feet or less, concentrated on a spot under the rear of the envelope. The guns from the ground had now ceased firing, silenced by Leefe Robinson's Very light and people had begun to gather in the streets of London and at their uncurtained windows to look at the drama being acted out 11,500 feet above their rooftops.

Hardly had the 21-year-old Lieutenant finished the new drum of Brock and Pomeroy than a small glow appeared at the spot fired at, and in a few seconds the whole of the rear of the *SL11* was ablaze. Lightened by the disposal of its 6,300 lb of bombs and also with its water ballast gone, the wooden-girder airship could gain no more height, only twist and roll in the probing beams from below, its motors racing with the measured, deep beat that had become a familiar sound to Londoners. The fire swiftly took hold of the remainder of the craft as Robinson emptied the third drum of his Lewis gun into it. The search-lights, now useless, were doused one by one, so that the Very lights which he fired off to relieve his feelings stood out the more brightly against the black backdrop of the sky. Quickly getting out of the way of the falling airship, the BE pilot heard a similar sound to an express train as the airship rushed past him, heeling over as it did so, as the white incandescence paled again to orange.

Meanwhile the same concentration of searchlights which had attracted Robinson had also caught the eye of Lieutenant Mackay so that he climbed and began to close in on the target. Being some distance off, this took him about 25 minutes and when the airship was almost in range he was astonished to see it suddenly fall in flames for no apparent reason since he was unaware that his colleague was also in the area. On the ground now, hundreds of people were gathered in the streets and a pedantic man is on record as having rushed into his home and grabbed a newspaper to read it outside, so great was the illumination given off by the doomed *SL11*, while at least one child of the time remembers her parents, clasped in each other's arms, executing a wild dance of joy round the front room of their home.

The crash came north of Enfield in the fields of Cuffley with a splintering of timbers and a sunset-like glow visible for miles and this continued for the next five hours watched at first only by a policeman and a farmer while the ammunition exploded all around from the intense heat. Next morning the fire brigade arrived, their horses steaming from the race up the hill and the last smouldering remains were doused. When the inevitable crowds arrived in ever increasing numbers and by every means of transport that wet and thundery Sunday there was no great skeleton to see, only thousands of feet of tangled wire from the bracing that was being rolled into gigantic masses. Later these were to be cut up and fashioned as bangles, brooches, pendants and pins, or just packeted in inch lengths and sold in huge numbers to souvenir-hungry Londoners in aid of the Red Cross.

Of the 16 members of *SL11*'s crew there were no survivors and their remains

The SL11 *souvenir booklet.*

were taken from the wreckage to the little iron chapel of St Andrew's that stood on the opposite side of the meadow from the Plough Inn where the inquest was held. The clearing up process revealed three Iron Crosses. One of these, together with a salvaged revolver, was presented to Leefe Robinson, now a national hero. He was awarded the Victoria Cross, while the Overseas Club was to offer him a gold watch, and numerous financial awards were to follow, among the first being that of £500 made by A.L. Oldfield of Grosvenor Place.

Three days after they had been killed, the crew of the Schutte-Lanz 11 was buried in the afternoon at Potter's Bar, the service being conducted by the Rev M. Handcock, assisted by the clergy of Potter's Bar, South Mimms and Northaw. Full military honours were accorded and the Japanese ash coffins were carried to their resting place by members of the Royal Flying Corps— Willy Schramm had returned to the land of his birth. Soon William Leefe Robinson, too, was dead. After the country had fêted him and every detail of his private life was publicised, including his engagement to pretty Joan Whipple, he was promoted Captain and became a Flight Commander with No 48 Squadron in France, in March 1917. Only two of the Bristol F2As which he was leading survived an interception on April 5. He was brought down by Vizefeldwebel Sebastian Festner of Jasta 11 flying an Albatross DIII.

A prisoner, Leefe Robinson attempted to escape from several of the eight camps where he was held until the last one was Holtzminden under the notorious Karl Niemeyer, a sadistic bully of a type which every country produces a number. It was largely due to the treatment he received here, including frequent periods of solitary confinement by 'Milwaukee Bill', Niemeyer's nickname from his American accent, that Robinson's health was broken; small wonder that when he was repatriated after the end of the war he fell victim to the epidemic of 'Spanish flu', as it was then called. After a period of delirium, when he imagined sentries with fixed bayonets round his bed, he died on December 31, after 17 days of freedom. On the day of the funeral an aeroplane dropped a cross of laurel leaves over his home at Stanmore and to the tune of Beethoven's Funeral March a bearer-party of RAF Captains bore him to his grave in the tiny cemetary of All Saints, Harrow Weald, at the junction of Uxbridge and Elms Roads.

In the years that followed, Londoners have not forgotten the greatest spectacle they had ever seen in their sky; until 1938 the vicar of Potter's Bar would hold an annual memorial service for the members of both *SL11*'s crew and that of *LZ31*. This was attended by officials of the German government and a plot of land over which the engagement was carried out was given for the erection of a memorial, the donor being Mrs J.M.B. Kidson of Nyn Park, Northaw. This took the form of an obilisk subscribed for by readers of the *Daily Express*. Among the inscriptions it bears one that reads: 'The award of the Victoria Cross to Captain Robinson was thus announced in the London Gazette of September, 1916, "For most conspicuous bravery". He attacked an enemy airship under circumstances of great difficulty and danger and sent it crashing to the ground as a flaming wreck. He had been in the air for more than two hours and had previously attacked another airship during his flight'.

Near the spot, when the tea room that had seen the inquest on the airship's crew was rebuilt as the Plough Inn during 1930, this portion became the saloon bar and when new gardens were laid out in 1955 they were named after Leefe Robinson. Later still his name was adopted by the pub almost opposite the

*An incendiary bomb of the type dropped
by airships. This one was dropped into the
Fire Station yard at Charlton Road in
September 1915 but it failed to ignite* (via
P. Lamb).

church where he is buried. This was opened in June 1954 at Brockhurst Corner,
one of its bars being called 'the Cockpit'.

The spate of souvenir postcards and booklets which followed what, with
splendid disregard for accuracy, Londoners called 'Zep Sunday' was to be
augmented within a month by others, the new victim being in fact a Zeppelin
airship. Where Leefe Robinson had taken off from Sutton's Farm, the victorious
pilot on this later occasion had risen from North Weald, an aerodrome about
eight miles to the north which had been established in 1916 and was to undergo
two programmes of enlargement, the first before 1938 and the second and more
intensive one between 1939 and 1945.

The pilot of the intercepting machine was Lieutenant W.J. Tempest who had
taken off at about 10 pm on October 1 1916. It was not long before he saw, held
in the concentration of the beams of a number of searchlights, an airship which
the press of the period later decided was a 'Super-Zeppelin'. What he did
not know was that the airship had on board that night none other than Kapitan-
leutnant Heinrich Mathy, the tall, 33-year-old commander who was hero-
worshipped by his crew and the German public. The former looked up to him as
a cool tactician who, despite the demands of service life, managed to combine it
with a happy marriage.

A member of an old Baden family, Henrich had entered the Imperial German
Navy 16 years before and had not joined the airship service until 1913 when he
had gained experience in both *L1* and *L2*. However, this had been short lived for

he had gone back to torpedo boats until January 1915 when he had taken command of *L9* before going on at *L13*. On this particular night he had behind him experience gained on more than 100 trips in Naval airships. On this October evening the distance that separated the aeroplane and the airship was no more than 15 miles and, as Tempest opened the throttle to close the gap, the anti-aircraft gunners from the capital's defences were throwing up 'a very inferno' of fire. The distance between the pair had been reduced to no more than five miles at an altitude of 15,000 feet when things started to go wrong for the pressure pump broke down so that Lieutenant Tempest had to maintain the petrol flow by the exhausting use of the hand pump, a stratagem that caused some distress in the cold and at that height.

In order to lighten the Zeppelin's load its bombs, (30 of them high explosive and 26 incendiary) were at this moment dropped, falling for the most part on Cheshunt where the most spectacular part of the damage caused was the destruction of several acres of glasshouses devoted to tomatoes. By now, the ground fire had ceased although the searchlights continued to hold the raider, among them the powerful one that was mounted in the wooden gallery with its many 'companion-ways' which was such a feature of war-time Hyde Park Corner on top of the Apsley Gate.

It was now obvious to Tempest that the *L31* was quite capable of leaving the danger area by simply dropping her water ballast and rising out of range so he decided to take advantage of what little height he had and dive down on the dirigible, firing as he went. The first burst had little seeming effect, nor had the second, delivered as he passed under the massive tail, which towered over him, so the pilot thereupon slid under the rear of the envelope, as had Leefe Robinson, and put in a long burst. He had not finished when the Zeppelin began 'to go red inside like an enormous Chinese lantern'. A matter of seconds later a huge flame roared out of the front of the vessel as it shot up a further 200 feet into the air, and it must have been only moments later that it was noticed by an eye-witness who recalls the spectacle thus:

'My mother and I had been busy in my shop, and were now going out to sweep the pavement so, having put all lights out, we took up our brushes to begin. It was very dark and very quiet and there seemed only us about in the street. We had been sweeping but after a short time my mother touched my arm and pointed up at the sky where there was something most awe-inspiring, for there, a very long way off was a shape that looked like a huge ship, very high up and smothered in flames; a really wonderful but frightening sight. These tongues of fire were orange and yellow, blowing and billowing about, and all the while it was slowly descending until at last it was out of sight.

'I asked my mother, ought we to phone the fire or police station, but she assured me that although there was seemingly no one about but us it certainly hadn't gone unnoticed. "God grant there is nobody in whatever object it is", she added, and with sobering thoughts we went quietly in to bed. Strangely no one seemed to mention the incident, and it was some time much later that we heard it was the German airship, *L31*, coming down.'

After the Zeppelin had ended its involuntary rise it seemed to hang for a moment before it came roaring down, stern first in the manner of most doomed airships, so that Tempest had to throw his machine into a spin to escape as Mathy's vessel shot past its destroyer, 'roaring like a furnace'. The blazing wreckage fell at Potter's Bar amid what appeared from the air to be a shower of sparks where it burnt out, and those who arrived first on the scene found the body of a man lying on the turf which bore the imprint of its fall, indicating that he had jumped before the final crash. The man, seemingly unharmed except for a slight distortion of the face, was on the point of death. He wore a heavy overcoat and thick muffler over Naval uniform—it was Heinrich Mathy, tempted over England by the calm conditions which people were every night offering up prayers to be replaced by high winds unsuitable for airships.

When the inevitable sightseers trecked to the fields of Potter's Bar the following morning and tried to catch a glimpse of the twisted metal impaled on a gaunt tree beyond the line of armed men that guarded the wreck, there were those who rejoiced in their hearts at the death of the Zeppelin commander but there would have also been those who recalled the quotation that would be chosen as the epitaph for Fregattenkapitan Peter Strasser two years later, 'Who art thou that judgest another man's servant. To his own master he standeth or falleth'.

Yet despite events such as these the district of London north of the Thames is connected in the minds of the majority with the aerodrome at Hendon, an area completely rural in 1910 when the foundations of the later flying field were laid by Claud Grahame-White. In the company of his friends he was seeking a spot for his future flying school and it was important that this should be within easy access of central London. The cold of January was not making travel any easier one morning when he came upon a location which immediately struck him as ideal for, with the line of the railway serving the Midlands on its eastern boundary, was an open space of pasture land extending for several acres.

It was only a matter of months before the aeroplanes arrived and the tranquillity of the scene was gone for ever and one of the first to use the field was Louis Paulhan, a contender for the newly-offered *Daily Mail* prize of £10,000 for a flight from London to Manchester. What followed was to be remembered as one of the most exciting races in the history of aviation for the other flyer was Grahame-White himself who, only 12 months later, was to organise a military flying display at Hendon for the benefit of the Parliamentary Air Defence Committee. Up to this time, he had largely used a Blériot monoplane for his work but, having decided that the type was not suitable for cross-country work, he procured a similar machine to that which his challenger would use, an Henri Farman with a 50 hp Gnome motor. This he learned to handle at Mourmelon and was away early on a bright, frosty morning en route for Rugby.

With only another 93 miles to go, he took off again only to be forced down with engine trouble near Lichfield. While standing here the strong wind overturned the Farman, wrecking it sufficiently to call for a train to be chartered for its return to Hendon. Despite gusty weather, Louis Paulhan made a preliminary flight on the evening of April 27 and announced his intention to start. The aircraft used, despite being similar to Grahame-White's, was actually a later model with reduced lower wing span. For repairs, the older machine was at Wormwood Scrubbs but on hearing the news it was taken from its shed and was away one hour less a few minutes after the Frenchman's aircraft, at 6.31-pm.

At Roade, 60 miles distant, the two were so far apart that any question of waiting for dawn was foolish so Grahame-White, although being the less experienced pilot, took off from the small field at 2.45 am with the opposite boundary hedge illuminated by the lamps of a large number of cyclists who were asked to co-operate; the first night flight in history had begun. In fact it almost stopped immediately for the motor began to cough and miss until the pilot discovered that he had caught his sleeve in the switch. A very rough flight then followed, guided by the lamps of friends' cars below and the Farman became increasingly difficult to control in the gusts, several completely swinging the machine round. Landfall was made at Polesworth at 4.14 am. Meanwhile Paulhan took off from Lichfield at dawn and managed to reach Manchester first, declaring on landing that so roughly had the wind treated him, at one point almost flinging him from his seat, that he would not make such a flight again for '10 times £10,000'. This was the first cross-country flight in history.

The following October, Hendon was again in the news. On a Saturday at the beginning of the month, an informal inauguration took place of the new Blériot Flying School, the eight aeroplanes being at first accommodated in a single shed although two were later taken over. Pierre Prier was later one of the instructors and his arrival over Colindale caused intense excitement, for many of the villagers had not previously seen an aeroplane. Aviation events at Hendon now crowded after one another in large numbers including another race organised by the *Daily Mail* in 1911. This same year saw consternation among Grahame-White's fans when he crashed his machine, overturning it but escaping unhurt, while 1912 saw T.O.M. (later Sir Thomas) Sopwith win the first Aerial Derby in a Blériot.

By the time of the outbreak of the First World War, the atmosphere at Hendon had changed and it had become an aircraft manufacturing centre and trams ran up the Edgware Road to transport the workers; many of the buildings that still remain on both sides of the road date from this time. The best known of these was the works of The Aircraft Manufacturing Company (Airco)

A Blackburn Dart on display in the New Types Park at Hendon in 1922. This was the first production machine (H.E.J. Monk).

Above *Parnall Puffin at Hendon on June 24 1922. This was the second aircraft produced for an experimental contract of three* (H.E.J. Monk).

Below *The air-sea rescue launch at Hendon RAF Museum which was included in the work schedule of the Museum's Society of Friends.*

Below right *Fokker D VII on display in the RAF Museum, Hendon.*

with, as its chief designer, Geoffrey de Havilland. Naturally all this activity brought a large number of temporary residents to the Colindale district and the local people accommodated them by scores in the small houses that still stand in many of the older side streets. Among these was the popular Philippe Marty who used to have the use of his friend Hamel's Morane monoplane. Sadly his machine got into a spin in 1913 and the well-loved airman died in the subsequent crash.

Throughout the years of conflict, apart from the manufacture of aircraft, the tradition of training here was continued through the Royal Flying Corps Civilian School of Instruction. A vast number of service pilots, including James McCudden, later to be awarded the Victoria Cross, like Warneford, another pupil, were trained here. They would have been familiar with the gates of the Grahame-White Company, the manufacturing concern which produced thousands of airframes under sub-contract, and which still stand near to the present entrance to the RAF Museum.

With the return of peace, sports flying again took place at Hendon and also joy-riding in ex-Service types, usually the Avro 504; but the real post-1918 memory for many was the selection of this aerodrome for the staging of the RAF Tournament, later to be re-named the RAF Pageant, an event that was instituted in 1920 to help restore flagging Service morale in the face of the short-sighted economy cuts by the politicians of the day. At these, there was a subtle mixture of items which had become traditional—such as the destruction of an observation balloon, and the bombing of a set-piece with a parade of the latest types of military machine, by means of the New Types Park. The aerobatics were often liberally marked by trailing clouds of beautiful coloured smoke and mock interceptions always concluded with one of the 'enemy' at least going

Above *'Flown covers' are very popular with stamp collectors. Handing over a bag containing some which have just been flown over London is Sir Peter Masefield, watched by Wing Commander R. Stanford Tuck.*

Below *A little-noticed exhibit outside the RAF Museum at Hendon is this British 22,000 lb 'Grand Slam' bomb* (L.D. Sorrell).

down with a long, red plume of vapour behind him. Event 'A' which started the programme at 12 noon was always either the Reserve of Officers' Race or the Headquarters contest and the traditional finale, five hours later, was the 'Air Battle'. Between these there had been all the usual items plus some of the less common to-day. These included exhibitions of crazy flying, air drill and flying boat parades.

Hendon continued in service use throughout the Second World War although it lost the custody of the King's Flight which had grown up here from No 24 (Communications) Squadron based here before 1939. After the conclusion of hostilities there was to be no resumption of the old flying displays, the last of these had taken place in 1938, although fighters, removed at the beginning of the London blitz, returned in the form of No 604 Squadron's Spitfires. Ten years later all flying ceased in 1957 when it was deemed too small for modern machines, although the last aeroplane to land there (on June 18 1968) is the Beverley transport which still stands outside the RAF Museum. This is housed in one of the old First World War hangars which may be identified by their wooden Warren Truss girders. This was opened by HM the Queen on November 15 1972 and so popular has it proved that 600,000 visitors arrived in the first year, a fact that speaks for itself since a special journey must be made to see the exhibits for its situation is not one which allows the public, as one warder put it, 'to come in out of the rain.' There is also an unbelievably flourishing Society of Friends, some of whom have contributed to the preparation of certain exhibits.

Across the car park is the Battle of Britain Museum, financed at a cost of £2,000,000 and opened by HM the Queen Mother. To list any of the exhibits in this, or the main Museum, would be useless for the aircraft on show are displayed in rotation, only the largest and most popular items seemingly remaining longest. One of these is undoubtedly the Short Sunderland flying boat which was towed into the new hall after a long sojourn in the open. This probably caused consternation only to a family of London starlings which had nested in it during the previous summer and may have been planning to raise their next brood in an aeronautical atmosphere the following year!

Chapter 7

London's Surrey

Leave central London via Vauxhall Bridge Road which has its own associations with aviation, as already described, and it is only a short distance into the part of Surrey which forms a belt to the metropolis. Much of the surrounding area was, in living memory, entirely rural and even those portions lying south of Norbury and Streatham near to Thornton Heath at the beginning of the 20th century could boast the occasional mansion where peacocks strutted on immaculate lawns, set in a semi-pastoral landscape. In consequence there was a natural tendency for would-be flyers to emigrate south of the river Thames rather than in the opposite direction which was the more densely populated.

In the early 1800s ballooning had become a popular sport now that gas as a lifting agent had replaced the dangers of hot air but there were many of its supporters who believed that the art had arrived at a point where, although there were plainly many problems to be overcome in the field of aerial navigation, few of its exponents were doing much to solve them.

Among those who felt that something should be done to reduce the hazards of landing a balloon was Thomas Harris, a formal Naval officer who had been attracted to the new sport; feelings which he expressed candidly when he wrote that 'the Science of Aerostation has recently fallen into much decay, and been the subject of ridicule through the total want of invention'. What he had in mind was that at the time there existed no means whereby a balloon could be rapidly deflated on landing so that there was an ever-present risk of the basket with its unfortunate occupants being dragged across the countryside, powerless to do anything about it, a problem which was later to be solved by the ripping panel. Harris' solution was to fit in the crown of the envelope a special valve consisting really of two valves, one operating within the other. For normal use the smaller, inner, one was opened by means of a cord, while quick deflation was achieved by pulling both this and a second cord so that a large quantity of gas escaped through the double valve in a very short time.

Although Harris was by no means a professional aeronaut it seems probable that he had aspirations in this direction for September 1823 had found him as assistant to George Graham when he flew from London to Rochester, Kent. It was this voyage that really fired his enthusiasm to construct a balloon of his own and this, a magnificent edifice of blue and yellow, was completed and named *The Royal George*, evidently in honour of the fourth Hanoverian king of that name who had only been on the throne four years.

Fitted with the new valve, this balloon was placed on public exhibition at the Royal Tennis Court which was at the southern end of Great Windmill Street across the road from the Haymarket in which district contemporary sources describe it, and here it attracted much attention in the spring of 1824. But Thomas Harris was anxious to take his new vessel on its maiden flight and soon announced that this was to be carried out on May 25 when he would rise from a tavern in City Road, the actual venue being *The Eagle* of nursery rhyme fame. On the appointed morning Harris arrived in the company of Miss Stocks, an 18-year-old whom, despite the voyage being in the nature of a test flight, he had promised to take aloft.

The Royal George rose without incident and, seized by the prevailing wind, began to drift southwards, a fact which would have caused no consternation to the pair in the basket since the open country under them was to be welcomed, being preferable to the Essex marshes towards which the south-westerly wind usually wafted balloons rising from London. They had been in the air for some little time when the two noted, slightly south-west of the little town of Croydon, a high, open space known as Dobbins Hill, probably what is now termed Duppas Hill and they decided to land there. Harris found no difficulty in the operation of his new device and valved gas in order to descend without difficulty. *The Royal George* was almost down when the first discovery to mar the bright May morning was made, the grapnel had been forgotten so the balloon was allowed to rise again and made off towards the neighbouring parish of Beddington. The journey was one of only a few minutes but was full of incident for the next expanse of open parkland which seemed to offer an inviting spot for landfall was the deer park near the parish church which stood beside the ancient Carew Manor, once the home of Henry VIII's Master of the Horse.

The girl's puzzlement at the sight of Harris seemingly trying to disentangle the two valve cords turned to alarm as the balloon began to plummet earthwards as he cried out 'Good God protect me!'. At the same moment a gamekeeper in Beddington Park reported a blue and yellow balloon falling at 'a fearful speed' and, before anyone could reach the spot, it struck the ground with a sickening crash, tearing a great limb from an oak tree as it did so. When help arrived, Miss Stocks was found semi-conscious and badly injured on the ground groaning, 'Oh Lord, where am I, have I fallen out of the car?', while Thomas Harris was found under the basket, killed instantly by the force of the impact, a fact that seems to support the French accounts which stated how, at the moment before impact, Harris had leapt out to lighten the load and save the girl. If this is so, the sacrifice was not in vain for she was to make a complete recovery and give evidence at the inquest. It was here that Graham suggested the wrong cord had been pulled so that the major one was opened in error, and at the time this theory was accepted by the majority. In the light of modern research it seems more likely that Harris made the mistake of fastening the line from the larger valve to the rim of the basket with little slack so that the loss of gas (made in the attempt to land a short time before) would cause the car to drop relative to the envelope and in turn increase the tension on the cord from the basket that the girl had meant when she spoke of tangled lines.

The Hodgson Collection still contains an aquatint after a drawing made on the spot in Beddington park by T.T. Dales showing the death of Thomas Harris. The style is reminiscent of Rousseau. It was to be further 15 years before the ripping panel was invented, a device that did the same work as the double valve but without the attendant danger. The inventor was John Wise who first used it on April 27 1839.

Already stated has been the fact that only a few hundred yards from the site of this 19th century aerial tragedy stands the ancient Church of St Mary the Virgin dating in part from the 13th century so that it is an unexpected repository of aviation lore. Along the wall of the old churchyard at its eastern boundary may be seen, secured by iron bands, a small collection of simple wooden memorial crosses; these date from the period of the First World War and were originally used as grave markers for the burials of British servicemen in France. When the cemetries on the continent were properly laid out and permanent headstones erected by the Imperial War Graves Commission, the wooden memorials were offered to the families of the fallen who frequently had them deposited in the care of their local church: there is an especially fine collection of these in Salisbury Cathedral, one of these at Beddington is to a Flying Corps Officer. 'In loving memory', reads the carved inscription, 'of my brother, Lieutenant Philip Mighell, 9th East Surreys, attached to Royal Flying Corps. Died of wounds at Agnes-Les-Duisans 12.10.17'.

Nearby, with a touch of macabre irony, is the grave of this officer's parents. His father, another Philip, was a victim of the total war that aviation made possible for he was killed on Wednesday, September 25 1940, when his home at The Horseshoe in Coulsdon, a small town close at hand, was bombed during the Battle of Britain. Life is once more given to these shades by the gentleman who remembers Mighell Senior (a name locally pronounced 'Mile'). Each Sunday he leads his family to worship in a long line up the aisle, children in descending order of age, with his wife Martha bringing up the rear—shadows of a vanished age.

Until the end of the 1960s this acre also contained another interesting relic of the aviation of the past in the form of a propeller cross. These were often put up over the graves of dead comrades by RFC squadrons and consisted of nothing more than a four-blade wooden airscrew with three arms cut off shorter than the fourth. At one time these were not uncommon in France. In England they were less easily found but here was one only 14 miles or so from London. This particular propeller cross was one from a Rolls Royce Falcon III motor, according to the regulation legend on the hub, and, unlike many, it had been protected by the addition of brass caps on the stubs of the truncated blades.

This was a memorial to Lieutenant Barkley-Winton of the Royal Flying Corps who died in 1918 aged 26. These facts were recorded on a plate which once occupied the centre of the cross, together with the information that he was buried at St Omer, France, and that Magdalene College, Cambridge, held a notice of his death, but by 1960 this plaque had become insecure and it was removed but is still preserved inside the church. The actual propeller cross, however, continued to stand despite the fact that, at the time the plate was taken elsewhere, the laminations were just beginning to break up. It is testimony to the workmanship of the makers and the durability of the wood that it stood so long, for the only attention it received was an annual coat of stain applied by Miss Hester Winton and her sister. Finally, however, the cross succumbed, not to the elements but to the ravages of wood beetles and Miss Hester requested the Rector to have the propeller taken down and burnt about 1968.

A memorial propeller-cross to Lieutenant Barkley-Winton, RFC, formerly to be seen in the churchyard of Beddington, Surrey (V.J. Garwood).

Less than a mile and a half east of this spot is another almost forgotten site which claimed an unwilling aerial voyager, the location being strictly situated in Waddon, part of the London Borough of Croydon, but some distance from the centre of the town. By this time the day of the balloon had passed in favour of the airship and in 1908, the world ablaze with the war of 1914-18 was still hidden in the mists of the future. England was also still an island, for just over a year was to pass before Louis Blériot was destined to cross the Channel.

The opening scenes of the saga were enacted at Wandsworth Gas works on the south bank of the River Thames not far from the railway station and within sight of the bridge that connects Hammersmith with York Road. Here, one magnificent May morning, Henry Spencer, the well-known Edwardian aeronaut, had taken his dirigible 80 feet-long airship to be charged with coal-gas before a trip to mark the first of the month which it was announced was to be 'overland, across London' culminating with a circuit of the dome of St Paul's Cathedral. As the inflation process proceeded it became evident that, despite the bright sunshine and the blue sky, a strong wind meant that a ground crew of 50 men was required to control the airship. Despite this, Spencer determined to go ahead and he and his passenger, J. Keaton, climbed into the car below the 30 foot diameter of the envelope. Having cast off, it became evident that even at a slight altitude the wind was going to prove a problem; even so, the pair made resolute attempts to head for the city as the little motor laboured to drive the dirigible along. Fifteen minutes later they found themselves in fact further from their goal than when they had set off, for whereas by this time they should have been across the river at least, they were south of the Gas Works and hovering over Wandsworth Town Hall. Here it was that the worst happened for the motor now gave up the struggle and failed completely so that the airship was to some extent at the mercy of the same prevailing wind that had played with so many balloons in earlier times, blowing them south-east. One says that they were not completely the plaything of the air currents because it was still possible, Spencer claimed, to have some degree of control over the vessel by means of the large rudder. It was in something of that spirit of resignation that marked all early aeronauts in lighter than air craft that the two noted Clapham Common pass underneath them.

Here it was not possible to land and, after a further period of drifting until they were about nine miles across country from the starting point at Wandsworth, the airship managed to make landfall in Factory Lane, Waddon, narrowly missing the row of gaunt trees that lined the road side. Naturally an unheralded visitor from the sky attracted a great crowd, particularly of young people on that sunny Friday, a fact that was reflected in the local press by no more than a short comment by a reader who remarked, 'I like that airship that wanted to go round St Pauls and found itself in Factory Lane, it would be so valuable in time of war . . .'. The two other chronicles of local events said nothing. Henry Spencer, on the other hand, was to become only too familiar with this part of London's Surrey and one suspects dreading the mention of it for, two years later, a capricious fate dumped him unceremoniously back there, in another month of May on the day that Parliament re-assembled.

W.S. Gilbert of *Mikado* and *Bab Ballad* fame once jested that he would chain himself to the railings of Westminster Lying-in Hospital and cry 'Babes for Men!' Had he done so it would have probably caused only a little more surprise than the sight of a Spencer airship when it took off from Hendon with a large white panel on the side of its envelope demanding in black capitals several feet high, 'Votes for Women'. To further the suffragette cause the passenger on this occasion was Miss Muriel Matthers and her instructions were to scatter a hundredweight of leaflets urging female sufferage over London while the airship progressed at a height of 3,000 feet. All went well, and the papers were well distributed but with the wind from the north-west it proved impossible to

de Havilland Moth G-ABEV Joan *was used for the re-enactment of Amy Johnson's pioneer flight in* Jason. *The replica restoration is seen here in the spring of 1980.*

An early airliner was Instone's Vimy Commercial, City of London, *seen here being loaded.*

turn back and instead the dirigible was carried on, over Croydon and Purley to Coulsdon, beyond where the trip ended somewhat inauspiciously when the airship descended to land over the open country and succeeded in crashing into a tree.

The spot was only a very short distance across country from the area that was to witness one of the London area's earliest attempts at flight. This was by aeroplane in a district which was then completely rural for, in 1910, the urban sprawl, and even that was broken in places, faded out altogether west of a line drawn roughly southwards from Catford, Caterham being the only built-up area of any note beyond Croydon. It therefore follows that Addington was a minute rural community, no more than a hamlet in fact, and it was here, up Spout Hill that a summer's afternoon found three young men pushing their bicycles. Two of these youths lived in Beckenham just over the county border in Kent at their father's home, 'Eastwell', No 4, Westgate Road. They were Arthur and Sidney Sippé. The latter (four years later and commissioned in the Royal Navy as a Flight Lieutenant) was destined to be one of the pilots flying Avro 504 machines which carried out the audacious raid on the German airship sheds at Friedrichshafen on Lake Constance as early as November 13, making it only the third British bombing attack of the First World War and probably the most brilliantly planned. The other member of the party, now breathing somewhat heavily as they neared the top of the hill, was a friend of the two brothers, James Jensen.

The intention of the trio was to seek an interview with the local farmer, a Mr

Still, and when they had been greeted by the cheery man they told him over the proffered glasses of sherry that they were building an aeroplane. They hoped to gain his permission to erect a shed in one of his fields to assemble the parts and to fly from the land on Castle Hill, the venue today being marked by Castle Hill Avenue. The brothers finished their request by assuring the farmer that this was no hare-brained scheme but one properly financed and devised since they had just completed their training as engineering apprentices in Manchester at the Westinghouse Works.

The young men's surprise was only equalled by their pleasure when permission was granted and shortly afterwards the parts of the semi-completed aeroplane were brought to the new shed that had been quickly put up. Some of the components had been specially manufactured in Manchester for the three had designed the machine almost a year before, a monoplane with a braced, steel tube fuselage and wooden wings although the rear of the body was no more than a boom formed to carry the tail unit on a single bamboo pole.

The shed where the strangely-shaped parts were assembled naturally became the focus of attention for the little local population and there were those among it who claimed that the work was already complete and the maiden flight had already taken place—at night! When the aeroplane was in fact finished and rolled out it proved of unusual interest, perhaps the feature most quickly remarked upon being the fact that the pilot sat low down under wings over which was mounted the 40 hp French in-line motor. The undercarriage was sprung after the manner of the Blériot monoplane with sliding steel tubes but the landing shocks were absorbed not by bungee but with helical springs. Probably none of those sightseers who stood about that day or suddenly found urgent work to do on Castle Hill realised that they were looking at the first machine in the world to take advantage of welded steel tube in its construction.

The process of warming up the engine was a long and dangerous one as it called for the inlet pipe to be wrapped in asbestos, doused in petrol and then set alight, and the efforts were finally rewarded when the motor sprang into roaring life. The monoplane's maiden flight was scheduled for April 24 1910, 12 days before the death of King Edward VIII, and a programme of intensive adjustment and experiment was carried out whenever time permitted, and when all controls were pronounced perfect, the wings were once more fitted as the strange-looking aeroplane stood again in the open air.

Word had, of course, gone round the hamlet with lightning speed and a good-sized crowd had assembled to watch the monoplane being wheeled somewhat unsteadily across the field while Sidney Sippé, accorded the somewhat doubtful honour of being test-pilot, sat at the controls and tried to look unconcerned. Then all was ready, the great moment had arrived and to the accompaniment of a rising note from the engine the machine began to rush with increasing speed across the grass, straight at the uncomfortably thick hedge at the far end. For a moment it seemed as if the flying machine would rise and clear the obstacle for few of the assembled watchers failed to claim that a distinct gap could be seen between the grass and the bicycle wheels. Whether or not this is true, the pilot realised that something must be done quickly and he closed the throttle. Alas he was not swift enough and disaster was unavoidable. It was not the hedge that did the damage but the ditch which ran at its foot, into this the wheels bucked, stuck fast, and the tail, trying to overtake the nose, only succeeded in rearing the whole machine up on its front as one of the wheels collapsed.

The first aeroplane to have a braced, steel-tube fuselage (via Croydon Advertiser Group of Newspapers).

Repairs were called for to both pilot and plane, the former had sustained the loss of a piece of flesh from one thigh, a broken nose and bruises; the latter wanted 'adjustments'. The nature of these were not particularly complex but they were expensive and the cash flow was now somewhat depleted. It seemed to answer all the three young men's problems when a visitor, Charles Lane the racing motorist, called at the Sippé home. He was, it transpired, starting a flying school at Oxford for the undergraduates and was seeking aeroplanes for ground instruction, so that naturally he had to ask if the Castle Hill machine could fly. At this point Sidney Sippé allowed his pride to exceed his better judgement when he answered with some strength that the machine was certainly capable of flying. 'It's no good to me then', said Lane, rising to go, and once again he explained that he only wanted machines that could be guaranteed not to depart from Mother earth. The inner struggle which Sippé suffered was short but fierce, 'All right', he finally blurted out, 'I'll make some adjustments and guarantee that it won't fly'. A bargain was struck, the 'adjustments' were made and it was no more than a few days before a sad little party of young men gathered to bid farewell to their flying machine as it was towed away to a new career at Oxford.

Strangely, nearly the whole of the continuing tale of the immediate surrounding district and the world of aviation is concerned with the war that followed the last halcyon English days, for even the first purpose-built London airport that sprang up in the midst of the quiet community owed its beginnings to the First World War. Undoubtedly the most spectacular of these was the appearance of 'the tram'. This was a specially constructed vehicle which, as part

Today's appearance of the old tram depot at Purley.

of the London defence system, would nightly emerge from the depot near to Purley in the Brighton Road and trundle its way to the limit of the line near to the ornamental fountain perhaps a mile distant. It almost marked the limit of the West Sub-Command which had its headquarters in the middle of Putney Heath, the actual operations room being situated on the top of the underground Metropolitan Waterworks and to this were connected all the 19 gun stations with their supporting 36 searchlights and 38 observation posts.

Strictly speaking the light at Banstead in Surrey was marginally further south than that at Purley but the former was a fixed light whereas the latter was fairly unusual in being mounted on the open upper deck of a tram. This vehicle was painted a sombre grey and had the windows of the lower deck boarded over, with the actual light, when not in use, hooded by a canvas cover. In order that the Royal Engineer crew which manned it might communicate with the local gun, an Army Field Telephone was carried. This could be plugged into sockets provided by the Post Office on every alternate lamp standard along the route with the final and most used one being at the terminus at the junction of Russell Hill and Godstone Road and it was here that the trolley arm was moved round with the aid of a long bamboo pole in preparation for the return journey.

There were nine vehicles of this type ordered for the London Defence System, six being for use in the Metropolitan area, two at Ilford and the single Purley example which was termed as an Advance Light for the local gun. This was an arrangement typical of the anti-aircraft precautions of the day although the type of gun varied. In some districts it was the practice for these weapons to be mounted on motorised chassis at the Kenwood House headquarters and from here to turn out in the manner of fire engines when warning of approaching airships was received. Londoners might be treated to the sight of several of these vehicles dashing across London with sirens wailing as they made their way to prepared sites in the anticipated path of a raider.

However, the gun served by the Purley light was of the fixed variety, in fact a 75 mm Quick Firer of French design and this was sited some four miles distant in the more densely-populated area of Thornton Heath. In those days the houses in Gonville Road terminated some two hundred yards short of the end of the road so that there was ample space for the gun at this point with a clear area to the north over the then 51 acres of open land. Due to the note of this gun when fired it was locally known as the 'Cough Drop'.

Each of these guns had also a nearer searchlight termed a Fighting Light, these being manned by six to eight men according to the type of generator, with an NCO in charge. The one for this particular gun was of the fixed variety in a field adjoining Aurelia Road and what was to become Shamrock Road, and the concrete base is still there under the buildings of the industrial organisation that now occupies the site as was the Army hut of the First World War vintage until it was demolished in 1977. As an illustration of the distances involved in this typical layout, the fixed light was scarcely a quarter of a mile distant from its gun while the unit was supported by a pair of Observation Posts, each on open areas about two miles respectively to the north and south of the gun.

As the Defence System became better organised, the mobile Advance Lights were dispensed with and that at Purley was now fixed not far from the tram terminus behind the hedges that then marked the end of Kingsdown Avenue where three private garages now stand almost facing the present Mount Park Avenue. It remained here until abandoned after the Armistice of 1918 when it became a rusting plaything for children, only to be finally removed by the Army.

Guns, too, were replaced by new weapons of greater capacity and, on a concrete base in the seven acres of fields bounded by Canterbury, Priory and Mitcham Roads, all built over in the 1920s when such areas as Onslow Road were laid out, a new gun was set up, soon to be locally called 'Big Bertha'. These weapons were usually manned by the Royal Garrison Artillery with 16 to 18 men under an officer assisted by a sergeant and corporal. Although to give details of their locations having discussed a typical one at some length would be wearisome, it is worth recording that an adjacent Beckenham gun, of similar type and calibre to the 'Cough Drop', was situated in Pickhurst Mead, Bromley, which shows a spacing of about four miles for the anti-aircraft defences.

To record the stories and locations of bomb damaged buildings of the Second World War would be unjustified and indeed not possible in a volume of this kind but it is felt that some mention is called for covering the earlier period if only for the fact that the historical repercussions were intense. In the vast number of cases, aerial attack was the first contact that a population had with the new science of aviation, and it was suffered by a generation whose attitudes and outlooks were those of a race bred in a centuries-old atmosphere of security, such as is only known by the people of an inviolate island. Small wonder that in the decades to follow 1918 an attitude grew up that was not to wither until the euphoria that followed the Battle of Britain, namely that the flyer was little more than a criminal, and a mad one at that.

The first air raids on London and its area were carried out by airships, mostly the large rigid designs to which Count von Zeppelin had given his name and from these, any form of bomb aiming was to all intents and purposes impossible. The injunction at one time issued by the German Kaiser that certain buildings were not to be attacked is only to show how ignorant his advisers were. London's Surrey, at this time scarcely awakened from its atmosphere of an Edwardian midsummer, was typical of the areas where small pockets of violence were scattered throughout. Often the dwellings were those at the time termed 'villas' for the reason, mainly, that the larger buildings were often sited

Above *Houses in Edridge Road, Croydon, following a bomb attack by a Zeppelin on October 13 1915* (Imperial War Museum).

Below *The same spot as it appears today.*

Above *Oval Road after the same incident in 1915* (Imperial War Museum).

Below *A photograph taken from the same spot in 1980.*

Left *Leonard Monteagle Barlow, MC and 2 Bars* (via Professor H.M. Barlow).

Below *'Ben Varne', 132 Onslow Gardens, Wallington, Surrey, the home of the Barlow family during the First World War.*

in the more open parts, so that the odds against their being his were immeasurably increased.

Typical was the attack mounted on the night of October 13 1915 when five airships crossed the coast of East Anglia, one of them, *L14*, with the redoubtable Captain Bocker in charge, missed London completely and instead wandered as far south as Shorncliffe and Winchelsea before turning to pass over Tunbridge Wells and Oxsted in Surrey. A little further and the first of 18 bombs were dropped, exploding with a brilliant flash in Edridge Road, Croydon, at a little past 11 pm. Witnesses at first mistook the sound of the approaching airship for that which was sometimes emitted by the gas street lamps of the period but as it approached it was likened to the noise of a heavy motor vehicle in another road. In fact, this first bomb of the stick flung a bed containing a sleeping mother and daughter into the street and a baby boy in the neighbouring house escaped injury also when the roof collapsed in such a way that he was protected from falling debris.

Other bombs were scattered over nearby roads and it was the sound of these explosions that caused a factory in the area to turn off all its lights. Some of the later bombs of the stick fell some short way off in Oval Road but the final one did not explode. It was as a result of this attack that regulations were introduced that restricted the illumination from street lamps in such a way that they were not visible from above and the glow on the pavements was curtailed to a circle reported to have been 'about three yards in diameter'.

A name strongly associated with military flying of the same period is that of Georges Constantinesco, the Rumanian engineer who perfected the gun synchronisation gear. This was based on the same principle as that which operates a pneumatic drill, so that a machine gun might be fired through the revolving arc of an airscrew. It is not generally known that he had family connections with Cheam, in Surrey, although as yet no memorial to him may be found in the locality.

Not so, however, a flyer of the period who is one of those, along with the names of aircraft and aero motors in the form of street names on the Roundshaw Estate, a few miles distant in Wallington. His name was Leonard Barlow. One of six children, he had been born at Highbury, north of London on June 5 1896 but his family had moved south and lived at 'Ben Varne', 132 Onslow Gardens, a large semi-detatched house. At the same time the war broke out, Leonard was an engineering student at Finsbury Polytechnic but he joined the Royal Flying Corps at the first opportunity aged 17; on his 18th birthday he was commissioned. His abilities as a pilot soon attracted the attention of General Trenchard so that he was soon posted to No 56 Squadron, formed with the intention of becoming a special, superior unit.

Here, alongside such exceptional men as Albert Ball, he quickly formed a friendship with James McCudden, later to be awarded the Victoria Cross and the latter makes several mentions of Leonard in his autobiography. Indeed, it was one of James' sisters who recalled for the writer the incident when he arrived unexpectedly late home on leave at their Kingston-on-Thames home in the company of Barlow and the pair wrapped themselves in blankets and spent the night on the floor of the warm kitchen rather than disturb the household.

Between April (when the squadron, including Barlow's 'B' Flight saw its first action) and October 1917, a total of 18 machines fell to Leonard's guns, the final one on the second of that month. The time in France, except for leave, was

Leonard Barlow beside his SE5 aircraft before leaving for France with No 56 Squadron, RFC. Note the quartered wheels (blue and white) and transparent hood on the aircraft (via Professor H.M. Barlow).

only broken by a period during June when the squadron was briefly recalled for Home Defence duties but only three false alerts resulted.

Perhaps the sortie for which he is best remembered is that flown on the afternoon of June 7, a special mission. The first action on that occasion took place at Bisseghem, identified by the Mineral Water Factory's advertisements, where he shot up the aircraft sheds. The next target was a train which he attacked twice from only 20 feet before leaving this for another, shot up from only 50 feet altitude although the smoke and steam made it impossible to see the results. At Wevelghem, troops in the streets and the railway station were shot up as well as a train stopped there. He next flew, still at only 50 feet, along the Menin road to Reckem aerodrome which was thoroughly straffed at times from an altitude below roof level, although the combat report makes no mention that a landing was made, as some historians claim. With a port elevator shot away by a ground gunner that he finally had to silence, Leonard then returned to base and the award of the Military Cross was gazetted on August 25. By the time of his posting back to England, for work as a test pilot in October, two Bars had been added to this decoration.

On Home Establishment, Leonard Barlow, now recommended for promotion to Captain, found the nearby Beddington (Croydon) Aerodrome most conveniently situated for visiting his family and a low approach over the roof of 'Ben Varne', with perhaps a wave to the two little boys who were playing in the garden across the road, would announce his arrival on the next available tram from the landing ground. Another story tells of the young girl next door who fainted as Leonard's wheels almost touched the roof as he came down to wave a greeting to her when she appeared at an upstairs window!

Sent to Martlesham Heath in the last week of 1917, Leonard Barlow was testing a Sopwith Dolphin, *C3779*, there on Tuesday, February 5, when the machine, which had already had an unsatisfactory engine replaced, inexplicably crashed from only 150 feet when the wing structure collapsed. The cause was never satisfactorily found, hardly surprising in the light of the fact that the

Leonard's brothers, Harold (left) and Donald, pose beside the road name-sign dedicated to their brother at Roundshaw, Wallington.

machine was totally destroyed by fire, but there were those who whispered of reversed controls.

His name is recorded on the local cenotaph—a memorial erected, it is believed, by the firm of Howe and White. In our own age they are usually associated with the maintenance and renovation of royal residences. But there is a greater interest for the aviation enthusiast in nearby Sutton, the London Borough now responsible for the administration of the whole area. Here in Manor Park the Royal Air Force badge on the memorial is of a very early type, one which it was later decided was heraldically incorrect in that it shows the RAF eagle within a circlet formed by a garter and buckle, a feature shared with

An early RAF badge on the war memorial at Sutton, Surrey.

A Messerschmitt 109E of III/ JG26, shot down on July 26 1940, is exhibited in connection with war savings at Croydon's Fairfield Hall car park shortly afterwards.

the RAF Memorial* on the Victoria Embankment in London which was unveiled in July 1923 and is described in another chapter.

The names of the men who are recorded on a nearby tablet represent flyers who could never have imagined, as they guided their primitive machines about the hostile skies, the developments that would take place in little more than two decades to come and a look of one such brings us full-cycle while preserving our chronology so that the chapter may end where it began, in Croydon. Here, a well-known landmark is the building which houses the Fairfield Halls; a household name for lovers of broadcast music. But, before this edifice was raised in the late 1950s, the site had been occupied for many years by an open-air car park. To here was brought, during the second half of August 1940, a Messerschmitt 109E-1 fighter to be exhibited to attract War Savings in connection with the local Spitfire Fund. The machine in question was Werk Nr *6296F*, the suffix indicating *Flugklar* after the repair of minor damage sustained at an earlier date. In fact on its last mission it had formed part of the *Stabschwarm* of III/JG26 on July 26 1940 with the Geschwader Technical Officer, Oberleutenant Werner Bartels, at the controls.

Just before 1 pm on that Wednesday afternoon the Messerschmitt had been set upon by Spitfires not far from Margate and, after receiving a serious head wound, the pilot had been obliged to make a forced landing at Northdown beside the Broadstairs to Margate railway line, a manoeuvre made the more difficult by having to avoid the high-tension electricity cables not far from the cornfield where the fighter eventually came to rest. The ironies of war meant that while Bartels was a prisoner receiving hospital treatment which would mean his repatriation three years later, his lightly-damaged Messerschmitt would be encouraging the contribution of sixpences (2½p) in neighbouring Surrey. A Lancaster bomber was later shown on the same spot.

*See also *Flight Royal* by Peter G. Cooksley, published by Patrick Stephens Ltd.

Chapter 8

The southern approaches

South of the River Thames there were two factors that historically contributed to the growth of the area from an aviation point of view. One of these, the chance one of the prevailing wind, has already been mentioned but the more important was the fact that at about the same time as the infant called flying was gaining strength, the southern approaches to the capital, except for a small tapering area that stretched almost as far as Croydon, were either widely scattered or largely non-existent 10 years after the turn of the century.

Further afield the position was even more convenient with the result that there sprang up such landing grounds as Chessington in 1922 and Hook, 12 years later. The same is true of the eastern side of the area although several of these aerodromes were of First World War origin; the short-lived South Ash, about 10 miles east of Biggin Hill, Kingsmill in use between 1917 and 1919, later enlarged to become West Malling 12 years later, with Leigh and Penshurst aerodromes, near Tonbridge. Other minor airfields have included at different times Redhill, established in 1935 and enlarged during the Second World War; Horne, exclusively in use during the period 1940-45 and the little-remembered Newchapel, closed after two years in 1919.

In addition to the actual presence of flying fields of one sort or another, the area also boasted, until June 1920, close in to London, its own airship station, HMS Crystal Palace and this station had also facilities for airframe instruction for heavier-than-air craft. Another RNAS balloon station was more distant at Roehampton, (No 1 Balloon Training Depot). The choice of the Crystal Palace for a balloon and airship training station was in fact more opportune than the Naval authorities realised for it had seen this type of craft there since Victorian times, one of the chief ascents of the period being that made on September 1 1862 by Coxwell and Glaisher plus no less than nine passengers in what was understandably described as 'a mammoth balloon'.

The whole surrounding area must have been very open at this time and the lack of development even as late as the First World War is shown by the adventure suffered by Cadet Woodman who took off from Hurlingham one Saturday morning with about 10 others in a balloon with their instructor. They had drifted across south London and were over Norbury when it was decided to land in the large estate surrounding Norfolk House on the top of Beaulah Hill. This they did and were at once the focus of attention of the gardeners and some of the maids from the big house all of whom hung onto the basket while the passengers disembarked, and to compensate for their weight a tarpaulin cover in

A seaplane construction class for Probationary Flight Sub-Lieutenants of the RNAS at Crystal Palace about 1917 (Imperial War Museum).

the car was filled with earth.

However, being a Saturday, when training exercises were regarded as something to be finished as soon as possible, the balloon had not been properly ballasted and when all were out except the unfortunate Mr Woodman, the balloon shot into the air with him still aboard! His last impression was the sight of the instructor's red, upturned face bawling instructions not to panic. He rose to 15,000 feet over Croydon still obeying the last injunction to hang on to the valve but however much gas he tried to lose it seemed to have little effect, not that this worried him much and he was meanwhile amusing himself by picking out a road where an uncle lived.

A few hailstorms later which made nonsense of the British Warm and kid gloves he was wearing changed his mood and as a precaution, lest he faint, he secured the valve rope round his chest so that, should he fall unconscious, gas would continue to be lost. Shortly afterwards he chanced to look at the barometer and realised that the balloon was indeed descending and at a most alarming rate, just how quickly he was soon to realise for the sand, jettisoned to arrest the fall, blew *upwards* into his face! He was now over open country and realised that there was no time to drop the sand by the regulation method of pouring as he was now at only 10,000 feet so he took a chance and flung the bags down intact where they burst in a field. Despite these measures the balloon continued to rush down and finally touched the earth on a railway line from which he bounced back into the air again and the involuntary journey continued to Chislehurst.

Now with some measure of control, Mr Woodman valved down and flung out the grapnel to catch in a May tree, but this he only succeeded in uprooting and trailing behind him! But at 30 mph and about 25 ft altitude he believed his troubles were almost over and in a final attempt to get down he pulled the rip panel and hit the ground with a bump that trapped him under the heavy basket so that for ten minutes he was a prisoner until a group of men released him. The balloon was folded up and put on a train for London where it would be collected by one of the Training Section lorries, probably from the Scrubbs, although other establishments existed at Richmond (No 2 Balloon Training Depot) and Kennington Oval, a Free Balloon Station that was to be closed in May 1918.

In the centre of the southern approaches to London in south-east England lies a group of three aerodromes, two of them now disused, and in their midst is the site of the tiny Addington landing field opened in 1933. The three major airfields here are Croydon, Biggin Hill and Kenley and it is the first of these that is the oldest for, established in 1915, it pre-dates the others by two years. In common with many of the flying fields about London, this was part of the London Defence System from December 1915 when some corn land at Waddon was pressed into service for a detatchment of that same unit which was to become No 39 Squadron of which Lieutenants Robinson and Tempest had been members. This was not all, for several training units were based here at the little aerodrome on the west side of the later, much enlarged airfield that was to be the London Airport of the 1930s. This fact was commemorated by the plaque unveiled by Lady Maud Hoare which used to grace the wall of the main entrance until it mysteriously vanished some years ago.

Here also was the National Aircraft Factory. To serve this a special railway line was laid down and the sheds of this became the well-known Aircraft Disposal Company from which so many pioneer flyers purchased their ex-service machines after the end of the First World War. This was a period when airmen stationed at Waddon, as it was sometimes termed, were billeted a few miles away in the little town of Wallington in the Gaiety Cinema that still

The bronze commemorative plaque which mysteriously vanished from Croydon Airport's terminal building. This is the only photograph known to exist of it.

Sopwith Snipes stacked in one of the Aircraft Disposal Company's sheds at Waddon, Surrey, after the end of the First World War.

'The Propeller' public house at Waddon, Surrey.

Detail of the Rotol airscrew at one time displayed outside the Off Licence.

stands, although only the rear is recognisable to those who remember that period of time.

This is only a part of the aviation influence which may be traced in all of the surrounding area. However, not all of it is directly connected with the presence of a flying field for the nearby town that has given its name to the airport, despite the fact that only about one third of the total area lay within the boundary, also suffered aerial attack before 1918.

Like every other part of England, this immediate district has some echo of the young men who were atttracted to the magic of flight, only to die in aerial combat so that they remain only as names on a war memorial. Thus there is some especial interest in a memorial to be seen in St Mark's Church, Woodcote, then a very rural district, that seems to record in one name that brief, late period when the Royal Air Force used army-type ranks so that L.W. Laming is described as a Private, RAF. The others are from the earlier day of the Royal Flying Corps being a Captain S.F. Browning and Second Lieutenants K.G. Cruickshank and N. Pearman.

With the enlargement of the Aerodrome and its elevation to the position of London Customs Airport a new terminal building was erected and after the closure of the centre in 1959, the stylised wings and globe that had for long adorned the entrance canopy were removed to the CAA building in Kingsway. Naturally the existence of the most modern airport in the world made great differences to the surrounding boroughs which were reflected in a number of ways. A motor works not far away, formerly Kirkways, now part of the Mann Egerton Group, described themselves as 'General, Automobile & Aeronautical Engineers', while the Borough where the greater part of the field's acreage lay, the former district of Beddington and Wallington, incorporated an aeroplane in its coat of arms as may still be seen on the Crown Court and Fire Station. On the other side of the boundary in the newly-laid Purley Way bypass, a public house was opened named The Propeller and between the end of the Second World War and the closure of the aerodrome a genuine Rotol airscrew was to be seen in use as a sign.

Short-lived was the airship mast that was used only twice, on both occasions by the *R33*, in 1921. This 140 foot high monster was erected at a cost of £25,000 and lasted only until September when it was hastily demolished leaving no trace. The official reason was that it was a hazard to aircraft but in fact it had been built in error on private land!

Not only scheduled flights took place from here but also charter trips and, in the early days, sports events. One of the best remembered is the Aerial Derby of August Bank Holiday, 1922, when the winner was Larry Carter flying a Bristol M1D, a variant of the monoplane scout of the war years. The top speed on this occasion was 108 mph. To a greeting from a crowd of 150,000, Charles Lindbergh came here in 1927 after his trans-Atlantic flight to Brussels and, with little ceremony, Amy Johnson left here for her lone flight to Australia in 1930.

Aircraft from every nation on earth used Croydon during the years to follow including Nazi Germany and it was to see the last of these leave, a Ju52/3m, 'Oswald Boelcke' that a little knot of people gathered to watch it go as peace ran out in 1939. Soon the RAF was back again in the form of No 615 Squadron's Gladiators as their home base, Kenley, was in the process of having concrete runways laid.

Above R33 *at the Croydon airship mooring mast, probably on July 14 1921, one of the only two occasions when the mast was used.*

Below *12.35 pm on Sunday, August 27 1939, and* D-AXOS, Oswald Boelcke, *the last Nazi aircraft to leave Croydon before the outbreak of the Second World War, prepares to take off.*

Above right *The damage to the old Croydon airport buildings, sustained in the first bombing attack on August 15 1940, may still be seen today. These are the doors of the former Air France hangar which have been pierced by bomb fragments.*

After the departure of the famous 'armada' of service aircraft for France in early 1940 the next major event came when Croydon was a fighter station with Hawker Hurricanes of 111 Squadron. Then, in the early evening of August 15 1940, the station and district suffered its first bombing attack and the hangars that no longer house aeroplanes still bear the damage sustained on that day when the 18 aircraft, mainly Messerschmitt 110s with some 109s of Erprobungsgruppe 210, attacked the aerodrome without any public warning having sounded, and succeeded in putting it out of full operational use for 48 hours. War-time events that followed did so with a pattern familiar throughout the south east; the destruction of a Blenheim in a crash not long after the outbreak of war was repeated on Friday, April 21 1944, when a Lancaster III, *ND582,* DXS of No 57 Squadron from East Kirby ploughed into Nos 55, 57 and 59 Lavender Vale at 2.20 am in an unsuccessful attempt to land, when returning damaged from a raid on La Chapelle. Two months later the first of the flying bombs came, a total of five exploding within the boundary of the airfield

After VJ-Day (Victory over Japan) civil services were gradually resumed, among the first operators being Railway Air Services still in their Service camouflage. Slowly it became clear that post-war air transport had no place for an aerodrome so small for fixed-wing aircraft and its forte became sports, charter and private flying until, on September 30 1959, the final closure came. The final scheduled departure was made at 6.35 pm by Morton Air Service's de Havilland Heron Series 1, *G-AOXL*, en route for Rotterdam. The last multi-engined machine airliner away was a Series 2 version of the same type and belonging to the same company, *G-AOXL*, flying to Glasgow, Captains G. Last and T.J. Gunn being the respective pilots. But the very last aircraft to leave was a Miles Gemini, that lifted off at 7.46 pm, only two hours 14 minutes before the official closure time. The historic machine was *G-AJWE* and Mr C.J. de Vere afterwards told the author that he was delayed in getting away as he and a friend

September 30 1959 and the very last aeroplane, piloted by Mr Christopher de Vere, waits to leave Croydon. The first users had been Service machines at the emergency landing ground almost half a century before. G-AJWE *was a Miles Gemini.*

had been burning (in effigy) Mr Harold Watkinson, the then Minister of Transport!

It then seemed that all was over and nothing remained of the district's flying connections except for such mute memorials as the severed Merebank, the ancient earthwork behind which smuggling is said to have been done in the very earliest days of the aerodrome, and of which a portion may still be made out at the side of the railway line. Another reminder is to be seen in the Croydon Parish Church with its fine example of the rare Battle of Britain Lace Panel. These reminders apart, the flying days seemed to have been relegated to history until May 1980 when the Spring Bank Holiday was marked by the re-awakening of the aerodrome for a single day's display. Residents who remembered the past days telephoned the local authority which had organised the event, and the Croydon Airport Society who co-operated, not to complain, but to congratulate all concerned on a brief return to former glories.

Had Nazi Germany invaded the British Isles in 1940 about six sections of gliders, supported by paratroops, it had been suggested, were to be dropped over the immediate area. A further half dozen companies of additional paratroops with one of troop-carrying gliders were to converge on neighbouring Sutton and the area north. That a largely non-industrial area should be picked out for attention in this manner becomes less surprising when the number of military aerodromes there are taken into consideration.

Of the same age as near-by Biggin Hill is the former fighter station at Kenley which was established as No 7 Aircraft Acceptance Park in 1917, its main duty being to prepare machines received from the manufacturers, for issue to operational units. It was summer when the first aeroplanes arrived and the new problems for the district posed by the presence of noisy contraptions whirling into the sky from part of the former common land caused an unusual item in the Station Commander's mail one morning. This was a letter from children of a local school requesting that there should be no flying on a certain afternoon lest the sound of the engines drowned the children's voices at a concert they were

Earlier, the last airliners had left, first the final scheduled flight, followed by Morton Airways' DH Heron Series 2, G-AOGO, seen here. Captain T.J. Gunn was at the controls.

Surprisingly aircraft were flying again from Croydon in the much-acclaimed air display of May 5 1980 including this de Havilland Tiger Moth floatplane.

The atmosphere of light aircraft was continued by this white and dark green Percival Mew Gull.

Left *To the collector even programmes become historic documents. This is the cover for the show at Kenley, Surrey, in 1937 when the highlight of the display was a formation of Gloster Gauntlets and a fly-past by a Fairey Hendon bomber.*

Below *The site today of the former emergency Operations Room for Kenley Aerodrome. This was once a butcher's shop in Godstone Road, Caterham Valley.*

presenting in aid of charity—there was no flying, by chance, and the school sent a letter of thanks.

As one of the RAF's oldest stations it is fitting that during the Peace Conference after the First World War, Kenley was the base of No 1 (Communications) Squadron flying a regular air service for delegates and mail between here and Paris. Not long after their DH 9as were replaced by the Avro 504s and Bristol F2B Fighters of No 24 Squadron. An officer of this unit was the first to die in the service of the RAF here, when Pilot Officer Matthew Charles Hayter was killed at Kenley on October 24 1922.

Communications work was replaced by that of a peace-time fighter squadron the following year when No 32 Squadron was reformed here on April 1 with Sopwith Snipes. This station remained its home for nearly 10 years with successive changes of equipment covering Grebes, Gamecocks, Siskins and ultimately Bristol Bulldogs. The first enabled them to appear as the stars of the display in connection with the Wembley Exhibition and the latter was the same type which had caused some excitement when one had landed with engine trouble on Streatham Common, seven miles distant, and turned over, avoiding a lady with a pram, in November 1933. However, before this, the station had become self-accounting and the General Strike had seen it become a distributions centre for official mail in 1926 and it was while based here that Pilot Officer Douglas Bader suffered the accident that cost him his legs.

With the expected expansion programme of the later pre-war years there followed a quiet time at Kenley between 1932 and 1934 during which the last of the old atmosphere was swept away in an extensive re-building programme and, following this, as an important part of the London Defence System, the Station became the base for Nos 3 and 17 Squadrons. It was during the time that followed that air displays became popular with large numbers of the visiting public. One of the best remembered of these was that marking Empire Air Day on May 29 1937 when 11 events were presented at Kenley in a programme that began at 2.15 pm, ending at 4.30 pm. There followed a half hour break for the performers before a repeat show was presented between 5 and 7.15 pm.

Yet despite all this there were more serious steps being taken such as the preparation of emergency Operations Rooms. This was done with little outward sign being visible on the aerodrome proper and was augmented by the requisitioning of an empty butcher's shop, Spice and Wallace, in Godstone Road, Caterham Valley. Here the premises were prepared for their new role with bays bricked up, etc, and telephone lines were connected from the main cable that the shop was almost directly over.

Best remembered of the early war years' attacks was that when the Station was devastated on August 18 1940. Mr Rignall, another butcher, but with his shop at Caterham-on-the-Hill, remembers the sight at about 1.20 pm on that day when Dornier 215s roared so low up the High Street that their black crosses and markings on the side could be clearly read. This was the day when No 615 Squadron flying Hurricane fighters, such as that preserved in the Science Museum, lost Sergeant Peter K. Walley when, at about 1.30 pm, his machine, *P2768*, fell to the guns of a Messerschmitt 109E over Morden Park Golf Course, which in those days presented a very rural appearance, quite different from that today. He had been one of the founder members of the Auxiliary unit, having previously been a fitter, and the only one from among the ground trades to be

MAXIME CINEMA, STREET

Thursday, October 2nd
Friday & Saturday

We proudly present:—

ANGELS
ONE FIVE

Cert. "U"

A Tribute to Britain's "Few" in their "Finest Hour"

Starring:—

Jack Hawkins   Michael Denison
Dulcie Gray

And the "Wings" that saved Britain and the World

Showing: Thursday and Friday 5.50 and 8.50

Saturday 5.35 and 8.50

Printed by E. C. Helliker & Sons Limited, Street.

sent for pilot training. He departed for this before the Squadron left for France and rejoined on July 1 1940 in his 20th year.

However, Kenley was also the base of other fighter Squadrons including No 253. Typical of their pilots was Flying Officer Alec A.G. Trueman, a Canadian who joined this unit three weeks after Peter Walley returned to No 615. During a ferocious fight over the aerodrome on Wednesday, September 4 1940, a prolonged burst of fire from a Nazi fighter sent his Hurricane *V6638* down out of control. The crash came some miles away in Banstead at about 9.45 am. The 26-year-old pilot was killed as the machine plummeted into No 18 Hillside Gardens, completely destroying the building and flinging debris into the garden and the house backing on to it so that it was easier for the owners, St Anne's Church, to have the remains pulled down and the house re-erected. Such was the force of the impact that the engine had to be recovered from a considerable depth.

As the war progressed other events overtook Kenley, one of the most historic being that when the Station Commander in February 1942 discovered the escaping *Scharnhorst, Gneisenau* and *Prinz Eugen* in the English Channel. This was Group Captain Victor Beamish who, a very short time later, vanished while flying from here. The following year, another famous name was associated with the little airfield when the Wing Commander, Flying, was J.E. 'Johnny' Johnson, DSO, DFC.

In 1944 the role of the Station took on a new significance when it became the operational headquarters for barrage balloons during the attacks by flying bombs and the next year Transport Command took over Kenley from the long-resident Fighter Command. The activity lessened at the aerodrome after 1945 only to be revived again in an unexpected form when, in 1951, the Station was chosen as the location for flying scenes and outside footage in connection with the first film about the Battle of Britain. This was the Associated British company's production, *Angels One Five*, so that Hawker Hurricanes again took off from its runways.

In 1980 the aerodrome was being described as a dormitory area for Officers and men working in the London Ministries and it was regarded as a Detachment of RAF Uxbridge although flying had not entirely ceased. On the one hand this was the care of the Air Training Corps in the form of No 615 Gliding School, the designation of which perpetuates an earlier association. The other is the occasional presentation of air displays. The first of these was that held to mark the Diamond Jubilee of the RAF on June 11 1978, followed by that on a wet and stormy day two years later; 40 years after the Battle of Britain. There also remains one association with the very beginnings of this aerodrome for on this site there was a hangar which was accidently destroyed by fire. It was a canvas one of the type known as Bessoneau, said to be one of only nine still in existence which, from its dull myrtle green shade, would appear to date from some time after June 1918.

Yet to a few, the little Surrey RAF Station which once shared responsibility for the defence of London and now is in use as a radio research centre and

Left *How cinemas advertised the first Battle of Britain film. All the filming was on location at RAF Kenley in 1951. Although it captured the atmosphere of 1940, the film was spoilt for some by portraying 'other ranks' as comics and showing a motley collection of flying kit.*

Above XA301, *a Slingsby Cadet TX3 of No 615 Gliding School at Kenley. These were later replaced by more modern aircraft.*
Left *Aviation pub signs are a study in themselves and in contrast to Croydon's unconventional one, this follows the more established pattern.*
Below left *The entrance to RAF Kenley in the 1980s. The Guard Room has been demolished at the left and the former Spitfire Gate Guardian has also vanished.*
Below *A Bessoneau canvas hangar erected in place of the permanent one destroyed by fire.*

extension to the Caterham depot of the Guards, will always be remembered. This is due to an amended sign, spotted by a senior officer visiting to watch an ENSA concert, which some service wag changed so that it read, not Kenley Aerodrome, but Kenley Hippodrome. It is from this date, some whisper, that official terminology began to speak of *airfields* instead!

Completing the triangle of aerodromes which dominate the southern approaches to London is Biggin Hill, dating from the same time as its smaller neighbour. First used for wireless experiments that were moved to the top of the North Downs from Joyce Green, which was too-often blanketed by fog, two officers from the latter had discovered the site in the autumn of 1916. Flying Bristol Fighters, the first operational unit here had been No 141 Squadron, intended to repel attacks by the growing number of Gothas and, when this need ceased, other experimental units concerned with the defence of London moved in, including the Instrument Design Establishment which migrated to Farnborough in 1922.

Until 1927 none other than No 56 Squadron had been the resident fighting unit but when this, together with a Night Flight, departed, the Station was ready for an extensive programme of reconstruction that was not finished until 1931 when Nos 23 and 32 Squadrons from Kenley became the new occupants. By 1936, 32 Squadron was alone until No 79's Gloster Gauntlets arrived in 1937, George VI's Coronation year.

The County of London Auxiliary Squadron, No 601, already mentioned in this narrative was in residence in time to exchange its Demons for Blenheims in 1938. By the outbreak of war, No 79 was in residence with others and on November 21 1939 this Squadron, brought into being by re-numbering 'B' Flight of No 32 Squadron, claimed the first victim for the Station in the form of a Dornier 17 sent down by one of their Hurricanes near Manston. After Dunkirk, which saw the Station sending a stream of patrols over the beleaguered beaches, the Battle of Britain brought 12 heavy attacks on the aerodrome. From two names of that era (Group Captain Grice and Felicity Hanbury) two roads are in commemoration today, both a few yards from the Guard Room on the opposite side.

On September 1 the Sector Operations Room was hit by a bomb from a Dornier 17Z and it was against such an eventuality that the Emergency Operations Room had earlier been established in a shop about a mile beyond South Camp, the site of the present Civil Airfield since 1959. At this point one is only a short distance from a block of shops across the road which at one time contained an establishment called the Teapot Gardens, identified by a large replica of that domestic vessel about four feet high, hanging outside. This was a constant draw to the young men at the Aerodrome who would from time to time 'borrow' it as a trophy. Alas, like the once young men who did this, the tea pot had succumbed to time and vanished in quite recent years although, painted red instead of its former gold, it certainly survived into the 1960s.

As the Second World War advanced, the nature of Biggin's use changed from a defensive to an offensive role, of which participation in the attack on Dieppe was one of the former in August 1942. However the years when the Station acted as a defender when Britain's back was to the wall had left their marks for those to find them. One such may be found at nearby Bromley Common where, at 26 Johnson Road, a Nazi bomber crashed into the front of the house with its entire bomb load still on board, in November 1940. While, a few doors away, a

Above *A present-day picture of the shop which served as Biggin Hill's Emergency Operations Room in 1940.*

Below *Houses in Johnson Road, Bromley Common, where a Heinkel bomber, with its bomb load still in the racks, plunged into part of No 26 (foreground) while one of the crew was suspended by his parachute from the chimney of No 14 during November 1940.*

Left *Individual memorials to air raid victims are rare. This one is on the wall of Elmers End Bus Garage.*

Below *The entrance to Church House, Bromley, where the Royal Observer Corps Operations Room was destroyed by fire on the night of April 16 1941.*

member of the Heinkel's crew hung from the roof suspended by his parachute lines.

That Biggin Hill was involved in the Blitz rather than the former daylight attacks with the turn of the year may still be seen behind Bromley Parish Church. Here the gardens have, among the flower beds, the unmistakable sign of a former house in the shape of an ornamental balustrade. This approximately marks the site of Church House that once stood here. The approach was originally marked by a large holly bush in the centre of an oval lawn and in an earlier age the carriages of visitors would come in past the entrance lodge and round the grass to the great central front door. Then, on the night of April 16 1941, this fine residence was set alight by incendiary bombs while the Royal Observer Corps plotters toiled in the Operations Room housed inside. Eventually they had to retire in orderly manner, taking their equipment with them and Church House had to be left to burn itself out. The next morning only part of one wall and the chimney stacks remained gaunt against the sky.

Biggin Hill took part in the operations following the D-Day landings but, about three weeks later, was completely evacuated. The reason for this was the opening of the flying bomb attacks. Balloon crews now took over the Station which suffered some hits from the missiles, as did the surrounding area. That this was true of the wider compass also is to be seen on the wall of Elmers End Bus Garage where a plaque records the names of those London Transport employees who were killed when a V-1 destroyed the old garage on July 18 1944; a tablet of this nature is fairly unique in the district.

It was September before Station Headquarters returned from its temporary home at Redhill and, although the use of the Station was as before, by fighters, it was not long before it was taken over by Transport Command which used it with increasing frequency until August 1946. No doubt the crews of these slower aircraft found just as convenient, as had the fighter pilots, the pub at Brastead known as the 'White Hart'. It was here that there remained in place for years a blackout shutter whereon many of 'the Few' had signed their names in chalk and in later years this momento had gained a frame and a protective plastic cover before it was finally removed to the RAF Museum at Hendon where it remains.

Now a part of Reserve Command, Biggin Hill was to undergo another change with a temporary return to Fighter control at the end of 1949 so that it was here that there took place the disbanding of Surrey's No 615 Squadron and of the city's No 600. The Station was now non-operational but there followed the establishment of the Civil Aerodrome and, until September 1980, the annual air displays kept alive the name in the minds of a vast cross-section of the public, as have the Air Fairs held each year in May since 1963. The present emphasis is now on civil aviation, particularly since the closure of Croydon.

There remain, however, some active service links with the former days while, as has been already said, North Camp is still in use by the RAF; the aerodrome also has St George's Chapel of Remembrance. The original one stood on a different site from 1943 until three years later when it was completely destroyed by fire but the Chaplain, the Rev C. King, who had conceived the idea in the first place, suggested the erection of a more permanent building since the first chapel had been not purpose built but a converted structure. This fact is reflected in the present design so that a certain severity and plain simplicity is incorporated in order to convey something of the atmosphere known by the

Above *A Douglas Dakota which had white, ultramarine and pale blue livery with red diamonds on the rudder, at the Biggin Hill Air Fair, 1970.*

Left *The altar of St George's Chapel of Remembrance.*

Above right *The last Battle of Britain display at Biggin Hill was marked by an assembly of four Spitfires and two Hawker Hurricanes.*

war-time personnel who were familiar with the first Station Chapel. Three architects combined their skills to obtain this effect, A. Beasley, W.S. Harper and G.A. Williams, the latter representing the former Air Ministry Works Directorate.

As the visitor turns right after passing the attendant Hurricane and Spitfire, although the last is not of war-time vintage, it is necessary to pass the foundation stone built into the wall on one's left. This was laid by Fighter Command's Commander-in-Chief during the Battle of Britain, Air Chief Marshal Lord Dowding, who was able to see the final dedication by the Lord Bishop of Rochester on November 10 1951. Inside attention is immediately drawn to the altar and its reredos and this has a very strong similarity to the war-time one, save that the number of names recorded thereon have now increased to total 453 pilots grouped against the badges of 52 squadrons based here at one time or another. At the time of its dedication the Chapel was in the care of Padre Vivian Symons and it was he who designed the unusual font, a gift by Lord Tedder on behalf of the Royal Air Forces Association. Between the three supporting fins beneath the bowl are three swords, each encircled by a gilt chain so that it represents the Station Crest. The blades of two of these are probably the oldest items in the Chapel since they first came from weapons forged for use in the Crusades, thus they must be in the region of at least seven hundred years old.

A few miles south of the Croydon, Kenley, Biggin Hill triangle lies Gatwick, less old than these three. It has been a civil aerodrome for all of its existence except for the period when it was taken over by Kenley for war use. The founder was Ronald Waters, a builder from Beckenham, who, as a resident of the old borough of Bromley and Beckenham, would have been familiar with Biggin Hill. He learned to fly at Croydon, thus forming another association with the famous trio. Waters at first set up his aviation business at the old Penshurst

Part of the former RAF Sector Station at Biggin Hill is now a civil airport and typical of the types using it is this 'Interflight' Aero Commander.

Aerodrome but, when this proved unsatisfactory, he turned his attention to the land that adjoined Gatwick Racecourse, itself opened as a replacement for the one at Croydon. The idea was that such a site would prove an alternative to Croydon Aerodrome when this was closed by fog for, on these occasions, diversions were made to Kenley to the surprise of passengers who suddenly found themselves landing amongst RAF machines.

With a certain amount of reluctance, no doubt occasioned in part by the tendency for the land to flood in heavy rain, a licence was granted with effect from August 1 1930. In point of fact the venture failed to make money for Waters and his partner, John Mockford, and the aerodrome was sold to Redwing Aircraft who established a flying school here in 1932. The following year it was sold again. The new owners were those who were responsible for the famous 'beehive' control tower and terminal building and at first named Horley Syndicate Limited they became Airports Limited at a later date.

British Airways transferred operations here in 1936. This was the year that the reconstructed airport was officially opened, on Saturday, June 6, by the Secretary of State for Air. More than 7,500 passengers arrived in four hours, their trains stopping at the old Titsey Green Station (re-named Gatwick Airport from June 1 but brought into service from the previous May 17). The opening of the aerodrome was marked by an air display. February of the following year was particularly wet and the drainage system at Gatwick was unable to cope so that a flooded British Airways pulled out in disgust.

Before and immediately after the Second World War, Gatwick is best remembered by the general public for its air displays, one of the final peacetime ones being held on June 25 1938 organised by the *Daily Express*. It was the same

Above *Aircraft at Gatwick, 1930. A Redwing Robin is in the foreground* (Pamlin Prints).
Below *Gatwick Airport in the 1980s.*

A Douglas Dakota of Hornton Airways Limited used for pleasure flights at Gatwick Air Display, 1949.

paper which presented the next display in 1949 when the final item—the Prince of Wales' Feathers drawn in red, white and blue smoke by three Spitfires—brought from the commentator the remark, when the red shade failed to function, that it was 'probably for political reasons'. Apart from events such as these Gatwick remained something of a white elephant until cabinet approval was given for the fresh development of the site as an alternative to Heathrow in 1952.

The result was the modern Gatwick Airport with such forward-looking proposals as that announced in 1978 when it was proposed that plans be formulated for the anticipated growth of passenger numbers to 25 million a year. Ideas such as this and the introduction of the 36 minute Railcar Link from Victoria was entirely asppropriate for an aerodrome which boasted retractable passenger walkways to its terminal in 1936. It all seemed a very long way from the brief excitement one war-time September day when a Junkers 88A-1 from 2/KG77, damaged by fighter attack and anti-aircraft fire, landed on the racecourse. One wonders if the crew of 3Z DK, *2142*, three of whom were captured wounded, had read the advertisements of the late 1930s which had described the area as 'exceptionally free from fog'.

Chapter 9

Westward flight

To the west of London the aviation connections are largely influenced by two aerodromes, Northolt and Heathrow. This is not to say that there are no others, equally important and even older while the contributions to the connections with flying in the area are just as far reaching. Perhaps chief among these is the old aerodrome at Hounslow, now no more than open land covered with coarse grass, although this was at one time the London Customs Aerodrome. It had been opened in 1916 as a service field as part of the defences. After the cessation of hostilities it saw the first operations of ex-military aircraft for early scheduled services. The first of these was started on August 25 1919 with a DH4 while DH9s were modified, some with enclosed accommodation for passengers who flew from London to Paris.

This was a regular service and was operated by Air Transport and Travel Limited at a time when passengers were issued with thick, fur-lined flying clothing before making a journey, extending to leather flying helmets even if they were to make the great adventure under the flimsy plastic covering that added little to the weight but made a contribution, however small, to the customers' sense of security! Perhaps from one point of view those who flew from Cricklewood were better off, for Handley Page Transport provided a certain limited degree of comfort since the machines that were used had, not long ago, been considered heavy bombers of the 0/400 and V/1500 type which, whatever their shortcomings, were sufficiently large for the provision of internal accommodation.

With the opening of Croydon as a civil airport, activities from Hounslow ceased in 1920 and the large board that flanked the wooden Customs shed was taken down. It had been much photographed as something of a novelty with its instructions to pilots and passengers set out in three languages. At the same time as Hounslow ceased to exist (only the year following that when the first four passengers had noted that it took only three-quarters of an hour to cross the Channel) Hanworth (Feltham) was opened only a mile distant, although this was always a private aerodrome and never operated scheduled services. The same is true of Brooklands, first opened in 1910 but roughly doubled in size during the 1939-45 war. This was the historic site of the Hawker assembly sheds and the aerodrome where a small silver monoplane lifted off the grass on November 6 1935: the maiden flight of *K5083*, the prototype Hawker Hurricane.

Privately owned, too, was the aerodrome at Wisley, just across the road from the more famous headquarters of the Royal Horticultural Society. From here many designs from the Vickers organisation first took to the air, but these did not include the predecessor of the Wellington, the Type 271, *K4049*, for this made its first test flight from Brooklands on June 15 1936 since Wisley, now disused but being examined as a potential executive jet aerodrome, only came into use in 1938. Just north of Woking, for five years from 1940, was Fairoaks only slightly longer in use than Smith's Lawn, closed after three years in 1945, while further north still was, between 1933 and 1936, the now-forgotten landing ground at Chalvey.

Before taking a look at the larger airfields in the area to the west of London there are two others worth a mention; Hawker's private aerodrome established at Langley in 1937 where, incidentally the Science Museum's Hurricane was prepared for exhibition and, almost six miles due north of this, Denham, sometimes known as Hawksridge, opened in 1935.

In an earlier chapter we have seen how the search made by Claud Grahame-White for a site which could be turned into an aerodrome for his new flying school, was finally settled when he chose the open pasture land that was to be seen from the region of Hendon Parish Church. He had taken a whole week to find what he regarded as the ideal spot and, during that time, looked at several possible venues in the rural districts which lay beyond London to the west, as well as the undeveloped area to the north. This was in 1910 and it is strange to note that among the sites examined and rejected in favour of Hendon were several acres of land near to the present site of Northolt Aerodrome.

The matter rested there for two more years until there was an organisation formed calling itself the National Aviation Company. This was to be responsible for the establishment of what it was proposed to call, Harrow Aerodrome, intended to lie north of the line of the Great Central Railway which had been opened five years before to serve an area almost completely agricultural.

So ambitious was the scheme that it was proposed to erect no less than 30 hangars plus a vast number of sports facilities, cricket and football pitches, etc, and also a number of aircraft, automobile and light engineering factories. The actual flying field was inspected early the same year and was pronounced ideal. That all these grandiose announcements sound like the assumed enthusiasm of estate agents is no surprise for this is just what the founders of the new company were—land speculators attempting to jump on the new aeronautical bandwagon just nine years after the Wright brothers had made their first tentative hop into the air at Kittyhawk in the United States. Naturally a project such as this can profit from every type of publicity it can gain and before long this was forthcoming from practically all the media of the day, newspapers and magazines.

What then happened to cut short all these ambitious hopes? History is silent on the reason, perhaps it was all a pipe-dream and there was insufficient cash. It is certain, however, that the exact site was never used and eventually became swallowed up in housing development such as that of the Field End Road Estate. Yet the area was undoubtedly suitable for aviation and in turn it attracted some enthusiasts who had a Blériot Monoplane there but which they never flew, and an airship experimenter who assembled a shed but had no dirigible to house in it.

The first use of Northolt, as we know the site today, was for training and as part of the search for suitable locations, Major W.S. Brackner (later Sir Sefton)

SE5a, F9053, *at Northolt in 1918 against an order placed with Vickers in July* (F.A.W. Mann).

reconnoitered the vicinity of the old 'Harrow Aerodrome' which lay across the railway from Downe Barn, Glebe and One Hundred Acre Farms. Due, it is said, to a misunderstood map, referred to by an assistant, it is the latter, in whole or part, which were requisitioned so that the work of preparation began here early in 1915. Those who looked for a suitable name, rejected the original proposition and also discarded the more logical one of Ruislip and took that of the railway station now known as South Ruislip but, in those days, Northolt Junction. The first flights seem to have taken place at the end of February.

Despite this it was not until the beginning of March that the first service unit arrived, in the form of a training squadron from Farnborough commanded by Major G.I. Carmicheal (later Group Captain). Even while instruction was going on the work of enlargement continued and one of the neighbouring farms was swallowed up in the process. It was now that the entry into the defence field took place with the ubiquitous BE2C once more pressed into use as a night fighter. For this work there were, as elsewhere, two of these at Northolt (later considered as a Home Defence Detachment of No 19 (Reserve) Squadron) which, in common with those at Sutton's Farm and Croydon, were to be redesignated No 39 Squadron on April 15 1916. It was the night of June 4 when the first alert was received to send the pair of converted two-seaters climbing into the summer darkness to seek out suspected enemy air activity over Kent. It was also the night when *L10* dropped bombs on Jarrow, aided by the lights of the shipyard, causing damage said to be in the region of £50,000 and killing 18 people.

As will have been observed already, at this time defence against any form of raider was very rough and ready and it was not until the autumn of the same year before an attempt was made to weld into a cohesive unit the scattered attempts of interception. The real test of these was not to take place until October 13 when airships *L11, L13, L14, L15* and *L16* all appeared over the coast of Kent and Essex and seemed to be heading for London, although the first, in fact, penetrated only as far as Norwich before turning out to sea once more. As far as Northolt was concerned the new arrangements were nothing short of a fiasco for the only BE2C which was serviceable could not take off as the visibility at ground level was too poor, although *L13* passed over Watford and Uxbridge en route for Guildford where it began to turn slowly towards the north-west.

The following winter was marked by a particularly wet period and this caused a multitude of problems on the new field since the drainage measures had up to now been primitive. The first attempt to rectify the problem took the form of spreading a considerable area with furnace ash and although this has since vanished from the surface it could still be made out in our present age from the contrasting shade of the grass when viewed from the air. The uses to which the airfield were put remained that of base for various types of training unit, at the time termed Reserve Squadrons, but even these saw active service from time to time. One took to the air on the morning of June 13 1917 when several rose to intercept a threatened raid by the Gothas of Kagohl 3 flying from their bases in Belgium.

Northolt-based interceptors on this occasion were Bristol F2B Fighters of No 35 Reserve Squadron now pressed into use in what was to prove probably the only occasion when aircraft from the Middlesex base were to meet the enemy in British skies. Meanwhile the raiders split into two groups and dropped their bombs on various parts of the docks and nearby areas, this being the first occasion in the First World War when Liverpool Street Station was damaged. It was a bomb dropped by one of the last Gothas over the main target area which sent up a public outcry and gave the popular press the chance to splash across their front pages the accusation 'Baby Killers!' for one of the last 50 kg bombs which was loosed chanced to fall on the London County Council's Infants School in Upper North Street, Poplar, killing a number of small children when the missile crashed through the roof. Later a memorial was raised on the spot with an inscription reading: 'In memory of 18 children who were killed* by a bomb dropped from a German aeroplane upon the LCC school, Upper North Street, Poplar, on the 13th June 1917. Alfred H. Warren OBE, Mayor. J. Buteux Skeggs, Town Clerk'. It was later claimed that another 30 with four adults were wounded. Luckier were those in a school in City Road where a bomb which penetrated five floors failed to explode.

It was when the raiders were re-grouping for the flight home and people were collecting their wits and gathering to remark on small details they had observed, such as the firing of a white flare by the leader before the bombing commenced, that the Bristol Fighter, piloted by Captain Cole-Hamilton and with Captain Keevil as gunner, made contact with the raiders. There was an exchange of fire and quite suddenly the rattle of the Lewis gun behind the pilot ceased as the weapon slewed to point harmlessly to the clouds and the Captain slumped in his

*Modern accounts mention only 16.

seat. On return to Northolt he was found to be dead, the only casualty in action against the enemy that the Station was to suffer in three years of war.

The Middlesex base was put to other purposes than training and home defence, however, and this included test flying for the products of the new Fairey factory in Hayes, but strangely there took place a little aircraft construction on the airfield itself. This was to be the erection of the Kennedy Giant, a massive machine by the standards of the day with a wing span of 142 feet that had been inspired by the Sikorsky *Ilya Mourometz* which had already flown in Russia. That this was so is no surprise since C.J.H. Mackenzie-Kennedy was a friend of the Russians and had been earlier involved in work on his large designs.

The construction was carried out in the open at Northolt as there was no accommodation large enough and the parts were brought from the works of the Fairey Company and that of the Gramophone Company Limited, now EMI. Throughout this period work was bedevilled by financial strictures since government support was withheld until the first flight had taken place. Meanwhile, Kennedy and his partners forged another aviation link with central London when the design office was opened at 102 Cromwell Road, South Kensington.

The erection took a very long time and it was November 1917 before the machine was ready for its maiden flight. By now, although it still resembled its Imperial Russian counterparts, it had undergone several major changes, including a shortening of the fuselage occasioned, it is said, when this broke during a shift of sites, and the fitting of a greatly enlarged rudder. An even greater change had also taken place in the power units for where the design clearly demanded four large motors there were only two tandem pairs of a mere 400 hp total each, all that a parsimonious Government would release so that the early British Salmson motors used meant that the monster was woefully underpowered.

Frank Courtney, then a Lieutenant instructor at the resident No 35 Training Squadron was prevailed upon to carry out the first flight. There were several attempts to take off and it became evident that the motors were unequal to the task that had called for the pull of two lorries and a 70-man ground crew when it was being shifted previous to the re-location of the wings. Determined not to be beaten, Courtney decided on a mildly desperate stratagem and the huge machine was taxied at full throttle down a slight slope away from the hangars and into a stiff breeze. After some reluctance, the main wheels unstuck and the machine rose into the air to make a hundred yard hop but so low that the tailskid remained trailing on the grass. At the bottom of the incline the Giant sank down again and settled into the soft ground. It never flew again although it remained (after removal to another part of the field) for many years as part of the local skyline until it finally rotted away.

During the years that followed the end of the First World War, although what had become by that time London's western RAF station, was never in danger of closure, its activities were sharply cut back but a measure of training continued to be carried on. The defence policy of the 1920s and after, demanded the presence of a fighter base to the west of London so that in this capacity Northolt was ideal. No 41 Squadron was reformed here in 1923 although they continued to fly ex-war-time Sopwith Snipes, later exchanged for Siskin III and IIIAs, and later for Bristol Bulldogs and Hawker Demons, the latter just before the squadron departed for Aden in 1935.

Fifteen months before this, on July 12, a new squadron had moved in with Gloster Gauntlets. This was No 111 which lived a fairly uneventful life here except for excursions to the annual air firing practice camp at Sutton Bridge. One of these (May 1936) was marred by the death of Pilot Officer Radice, lost when his machine dived into the River Nene while returning from the Holbeach range in poor conditions. The proximity of the base to London meant that it frequently enjoyed visits from a number of important people and, on July 8 of the same year, 'Treble One' was honoured by a visit from King Edward VIII, accompanied by the Duke of York (later King George VI). Also that year an inspection was carried out by Mohammed Ihsan, Chief of Staff of the Afghan Air Force, who was followed a week later by Air Marshal Dowding who presented the new Squadron Crest. Before this, No 111 had briefly left Northolt to combine with No 56 Squadron when the two carried out a fly-past when King Edward unveiled the war memorial on Vimy Ridge.

The next major event at Northolt was undoubtedly when No 111 was chosen to be re-equipped with the new Hurricane fighter, the first arriving in January 1938. From then on there was a frequent stream of visitors to the base to see the new wonder planes and this enthusiasm was increased after February 10. On this day the Commanding Officer, Squadron Leader J.W. Gillan, flew one of the new machines from Edinburgh to Northolt at 15,000 feet, averaging a speed of 408.75 mph. The press gave more prominence to the difficulties imposed by the thick cloud and ice formation than the strong following wind which had prevailed throughout the journey.

Northolt's geographical position meant that it could still be used for some aspects of training and typical is the occasion at the height of the Battle of Britain when No 43 Squadron, after three punishing weeks and losses in killed and wounded, was moved back here from Tangmere. The victory of the Battle meant an inevitable change in the pattern of air fighting which ceased to be so completely defensive with the introduction of *Rhubarb* and *Ramrod* operations. These brought several new squadrons to Northolt including a number of Polish origin, so that it is fitting that in Western Avenue there stands today the Polish Air Force Memorial. Situated as it is on the north east edge of the aerodrome its sharp simplicty stands out keenly and this may have been deliberately intended by the designer, M. Lubelski, who, as a Polish sculptor of note, was also responsible for the bronze eagle that surmounts the monument.

The idea of such a memorial was first suggested in the early part of 1943, just about the same point in time when the Polish units stationed nearby had just passed their numerical peak. A fund was started and from this the total cost was met of £8,172 7s 1d (£8,172.35½) in time for work to begin in May 1948, the site having been granted for a peppercorn rent over 999 years by the Middlesex County Council. The monument is constructed of York stone and bears the names of 1,241 Polish airmen killed on operations, just about half the total number of the 2,408 who perished, there being not enough room for a record of every one.

It was the Chief Chaplain of the Polish Air Force, the Rev Gogolinski, who suggested the quotation from II Timothy, IV 7 which is inscribed on the back of the memorial. This reads 'I have fought a good fight, I have finished my course, I have kept the Faith'. Since the cost was met from contributions both by members of the Polish Forces and the British public it is fitting that both should be represented at the unveiling ceremony carried out by Lord Tedder on

November 2 1948 at a ceremony attended by Augusy Zaleski, the President of the Polish Republic, who laid a wreath on behalf of its people. An annual ceremony of commemoration is still carried out on this spot by the Polish Air Force Association.

Two of the most historic occasions in which Northolt played a part occurred in 1945. The first of these was when, on July 25, Winston Churchill returned to the Station, flying back from the Potsdam Conference and it was from here that he made one of his final war-time journeys (on March 23) departing to visit Field Marshal Montgomery's Headquarters on the continent.

By this time the area of the field had been almost doubled from that available in the 1930s and this may have had something to do with the fact that, on March 1 of the following year, the Station was lent by the Air Ministry to the Ministry of Civil Aviation for use by British internal services and for flights to Europe. It was just over two years later that Northolt hit the headlines in a manner which every flying man deplores. It was July 4 1948 when an Avro York of RAF Transport Command was involved in a collision with a DC-6 of Scandanavian Airlines during conditions of reduced visibility. In this all 39 passengers were killed. One of the passengers in the Service machine was Sir Edward Gent, High Commissioner for the Federation of Malaya. There was a happier event for the press to record in 1950 for, on that date, it was announced that the former Airport Commandant, Air Vice-Marshal S.P. Simpson, OBE, DSO, was, from February 1, to become Divisional Controller of the MCA's South Eastern Division.

By now the period of loan to civil aviation was almost at an end and for only a short while longer would the ex-war-time Dakotas take-off, for example, to Paris or the Channel Islands, before 1954 would bring the closure of the airport for civil flying on October 31. It was then handed back to the RAF for continued use as a terminal for Service transport aircraft and, in the last decade, Hawker Siddeley HS 125s and Andovers became typical of the fixed wing aircraft, in addition to a widening variety of helicopters.

A Gazelle helicopter from RAF Northolt.

It is less than three miles across country from Heston (which sprang up in 1929 for a brief 19-year existence) to the aerodrome sometimes called, in earlier days, Harmondsworth. The former had enjoyed a short time in the glare of historical events when Prime Minister, Neville Chamberlain, came here with his 'peace in our time' message on the evening of Friday, September 30 1938. Nine years before the flying field at Harmondsworth had been known (if at all) by the general public, as the Richard Fairey Great West Aerodrome. From roughly this same district the brothers, Ross and Keith Smith, had taxied out on a bleak November morning, when the soft snow crunched under the wheels of the Vickers Vimy, to begin their epic flight to Australia.

That Harmondsworth was to become Heathrow, the official title until March 25 1946, no one at that time could have dreamed for it was to be 1942 before the real beginning of what we now know can be said to have sprung. Not that the

Above G-EAOU, *the Vickers Vimy used by Ross and Keith Smith* (Vickers).

Left *Northolt Station Badge with its motto* Aut Portare, Aut Pugnare, Prompti.

Right *Westwards' Britten-Norman Islander 2A,* G-AXHE, *used for the Heathrow-Gatwick Shuttle in 1969* (L.J. Dickson).

demands of civil flying were in any way to the forefront at that period. The first surveys were made to find a suitable spot for a large RAF transport base. Laying the runways began two years later and the work was still incomplete when, in 1945, the site was handed over to the civil authorities who were then considering plans for a new London Airport since it was evident that the former one at Croydon now had only a limited potential.

The runway pattern was still that typical of Service layout although still incomplete, and there was no form of accommodation for passengers so that a veritable village of brown canvas tents sprang up complete with duckboards laid as connecting paths and with fire buckets at the entrance to each. The whole resembled almost nothing except the site of a travelling circus which had pitched its tents about a modest, brick structure with a long pre-fabricated annexe. This, the only permanent structure to be seen, was the first control tower which was not replaced until nine years later in 1955.

Appropriately, the first take-off was made on New Year's Day 1946 when British South American Airways' *Star Light*, a 13-seat Lancastrian, converted from a Lancaster bomber, left on a survey flight across the South Atlantic. This Corporation was later to amalgamate with British Overseas Airways which was at that time using Hurn, near Bournemouth, for their scheduled services although they were later to move to Heathrow as were Qantas Empire Airways. It was once more a Lancastrian machine that was to inaugurate BOAC flights from here when six passengers left for Sydney, Australia, on May 27, by which time the name of the aerodrome had been officially changed to London Airport from March 25, although the term Heathrow has never completely faded out. Both of these last pre-date the formal opening for scheduled services on May 31 1946.

Above *A Tupolev airliner of the Soviet Union at London Airport* (L.J. Dickson).
Right *The Alcock and Brown Memorial at London Airport* (Vickers).

While events such as these were taking place the work of expansion and development was being continued and, on January 30 1947, the Report of the Layout Panel was published. The final form of the airport was to cover six and a quarter square miles with nine runways arranged in three parallel sets in three directions. South of the Bath Road there would be a total of six, the remaining three lying to the north. To accomplish such a vast building project it was envisaged that the work would be divided into three stages, the first of which was then complete and the second being at that time under way. Stage three was scheduled to begin in 1950.

At that time the passenger flow for a representative period of six months was in the region of 63,000 and it appeared that, although by 1953, the total would be in the region of one million, the figure would level off at about this point. Consequently it seemed that the best use was being made of the large flat area which was once part of the bed of the River Thames so that the siting of the new terminal building, and associated accommodation on a central island which was incapable of extension, seemed justified.

Plans for further enlargement received a set back in 1952 when the Minister of Transport and Civil Aviation, Mr A.T. Lennox-Boyd, announced in the Commons that the extension north of the Bath Road was not to be proceeded with as the additional amount of traffic which could be handled as a result would not justify the expenditure and disturbance. It was the same Minister who unveiled the Alcock and Brown Memorial two years later on June 15, the 35th anniversary of the original flight.

There were many historic events crowded into 1955 which could never have been envisaged only 15 years before when the first prototype of the Horsa troop

An early model of the Lockheed TriStar typifies the changes that London Airport has seen since the days when it was used by the Fairey Aviation Company.

glider had flown from the aerodrome on September 12 1941. Now, three days after Christmas, the de Havilland Comet III was to arrive there with Group Captain John Cunningham at the controls, at the end of a round the world flight. The last 3,350 miles from Montreal had been completed in a time of six hours and eight minutes but at an average speed of 548 mph—a vast contrast with the Lancastrian flights of the inaugural years when as many as 25 stops were called for on a flight to the Far East. Earlier, the vintage year had seen the transfer to London Airport of the Air Traffic Control Centre for South East England from Uxbridge. The operations began on October 28.

In addition to the opening of the new 127 foot-high control tower that year the new passenger terminal, now known as Terminal 2 and Queen's Building also came into operation. Six years later Terminal 3 was opened so that all passenger traffic was handled within the central complex area. Freight continued at this time to be taken care of on the north side and in other parts of the airport but, in 1968, a new Cargo Terminal was opened so that the area south west of Runway 5 now takes care of all cargo.

In 1977 (on December 16) there took place another historic and significant event at London Airport for Her Majesty The Queen opened a new Underground station on this day. At first it may seem that there is little unusual in such an event but this new Piccadilly Line stop was to be called Heathrow Central. It then became possible for passengers to have direct rail links with Inter-City trains serving Euston, with its connection via the Victoria Line at Green Park, and with St Pancras and King's Cross at Holborn, where no change is required unlike the Central Line connection to here from Liverpool Street.

A de Havilland Comet I, G-ALZK (British Airways).

From Heathrow Central it is possible to gain access to the three air terminals: Terminal 1 for British services to the continent and flights to Ireland; Terminal 2 served by airlines from overseas on short-haul work, European airlines and those of the Middle East; and Terminal 3 for inter-continental services. The Underground station has a number of tunnels radiating to each of these and the decor of the subways is sufficiently different to prevent confusion between them and the moving walkways were designed to permit the use of baggage trolleys and still not obstruct the way of other passengers so that the resultant tunnel width is 26 feet.

Of the original proposed nine runways only seven were ever to actually appear, the remaining two falling victim to the 1952 change of proposals and these have been further reduced to three, being the two parallel ones running east to west, No 1 alongside the Bath Road and No 5 across the south side near to the cargo area. The third runway is No 2. Shorter than the others, 7,734 feet as against 12,000 feet in length, although, at 150 feet broad it is the same as the others, this one is intended for emergency use or for employment in very strong cross-winds. This No 2 runway runs diagonally in a south-westerly direction and is a remnant of the original layout. The reason for this gradual reduction in the number of runways is the spread of the passenger terminals and the resultant encroachment on adjacent areas. The erection of fresh terminals at an airport are by no means the end of the matter since all must have their complement of piers and jetties.

At the time of its earliest use the original Terminal 2 was called the Europa Building and since it is the oldest of the passenger buildings in use it has fairly recently been the subject of a re-development programme which has succeeded

A North Star airliner of the former Trans-Canada Airways at London Airport, c 1954.

in changing its appearance almost beyond recognition. The incorporated improvements include a system of ramps to connect the check-in points on the ground floor with the upper level and a new coach station, office block and gate-room have been added. In addition the lounges, bars, restaurants and shops have been modernised and some enlargement has taken place for, although the introduction of the wide-bodied advanced jet liner has resulted in some easing of the problems associated with increasing aircraft movement, the reverse has been the effect on the number of passengers actually handled, in simple terms the total has roughly doubled.

Terminal 3 lies on the south-west side of the central complex and, after the departures building was opened, further improvements were incorporated including a new pier to serve the number of passengers disgorged by each visiting 'Jumbo' airliner. This is the Terminal from which operate all flights made by Concorde machines and the reason for this choice is that the two annexes are capable of handling 6,000 passengers per hour while there are also 40 immigration desks. No 3 is the latest terminal which, operating on two levels so that arrivals and departures are handled entirely separately, the building is some 6,000 feet long. As a further aid to speed and efficiency in coping with the terminal's throughput a sub-division of traffic is made so that domestic and overseas passengers are segregated.

Yet, despite the provisions for increasing traffic such as have been outlined the annual number of passengers passing through has already passed the total of 26 million, so that it is becoming increasingly imperative that another London airport be found. This also explains why Gatwick has undergone such a dramatic metamorphosis in a comparatively short space of time, with an annual increase in the region of 15 per cent, saturation point is little more than a few years off.

Several potential answers seem to present themselves, among them a Terminal 4 at London airport or a second one at Gatwick. Both will be necessary by the middle of the decade. Looking at the wider spectrum it became obvious, in the spring of 1979, that all the old sites would have to be re-examined; Stansted, Maplin, Cublington, Thurleigh and Nuthampstead, all except one having featured in the 1970 list of the Roskill Commission. Although there seemed to be a strong consensus of opinion in the country over a long period favouring Maplin this overlooked the enormity of the problem posed by its preparation in so short a time, leaving Stansted as the seeming obvious choice. Opposition groups inevitably and understandably sprang up and the essence of all their arguments was the environmental disturbance factor. They were agreed, however, that south-east England had no sites which were uninhabited and which would be capable of accepting an airport of some eight square miles in area. The best that could be achieved was the involvement of hundreds of people rather than thousands.

Whichever site is chosen it will probably never have a venue for a garden party or exhibit a famous aeroplane. Both certainly happened to London Airport which was chosen for the Royal Aeronautical Society's Garden Party on an appallingly wet Sunday in the early 1950s. The Supermarine S6A racing seaplane was also exhibited here from August 1957. We have already met this in work when N248, masquarading as the S6B *S1596*, was shown on the roof of the Passenger Building.

There is one last item of aeronautica to be found in the area roughly to the west of London which, although minor, is worth recording: this is the house in Arkwright Road, in Hampstead. That this was once the home of Sir Thomas Beecham the conductor is well-known, but how many are aware that in the days before 1918 this was the RFC Medical Examination Centre?

Chapter 10

Kent and the southern counties

Casting one's net over the southern counties of England beyond the area largely mentioned is sure to gather a large harvest, particularly in the county of Kent where the associations with aviation range from the large to small and, in the vicinity of Dover, gather particularly thickly. Aeronautics, like so many of man's other endeavours, have not been slow to make use of the fact that at this point the Channel is at its most narrow.

That these endeavours are many and varied is surely shown by the spread of these associations from the sands of Margate. These, in common with those at Brighton, were scheduled as a Category D3 airfield in 1919 to the spot where Franz von Werra had been shot down by John Webster of No 41 Squadron RAF flying a Spitfire out of Hornchurch. The exact location was, for the record, in one of the large fields on Winchet Hill's east side near Love's and Mannington's Farms at Curtisden Green. The date was Tuesday, September 5 1940, and the

An early postcard showing the open nature of the Blériot landing site at Dover.

Blériot Memorial Dover.

The inscription reads:

After making the first Channel flight by aeroplane, Louis Blériot landed at this spot on Sunday, 25th July 1909.

location is only a little distant from Maidstone. Not far from this lies the connection with the earlier conflict at the Parish Church of East Farleigh and Coxheath. Here may be found a propeller memorial to Captain Stephen Walter who was killed in action on July 31 1917 flying a DH5 of No 32 Squadron RFC. There is also a memorial window to this 20-year-old officer whose home was the Parsonage of this Parish.

Yet, despite associations such as these, one must turn to Dover for really concentrated nostalgia. Perhaps the earliest connections with the ability to navigate the heavens near here was established as early as 1785 when Dr Jeffries and Jean Pierre Blanchard left these shores in a hot air balloon on January 7. As the lifting agent cooled half way across the Channel, with a resultant loss of altitude, maps, instruments, books, anchors and even bottles of wine went overboard to lighten the vessel. Plainly this was not enough so that the voyagers' coats and trousers followed over the side, a stratagem which seemed to have saved the day, as they landed in a Picardy forest. Here they had to cling to the branches 'as naked almost as the tree' until ladders were brought to see them safely down after half an hour and later the town officials sent a six-horse wagon to collect the balloon for lodging in the principal church.

It was almost one hundred years later that another crossing was made which rivalled the earlier one in its bizarre nature for 'Colonel' Samuel Franklin Cody went across the 22-mile strait seated in a small canoe pulled by a kite so the motive power, if not the passenger, was airborne. History reserved a special place, however, for the next important experimenter who came in the other direction on the early Sunday morning of July 25 1909. This was, of course, Louis Blériot, still suffering from the burns he had sustained to the foot, brought about by incautious contact with a hot exhaust. This landing was a heavy one which damaged the undercarriage and broke the airscrew of which

The same location as it now appears.

... ER MAKING THE FIRST CHANNEL FLIGHT
BY AE OPLANE
LOUIS B ÉRIOT
LANDED AT THIS SPOT
ON SUNDAY 25TH JULY 1909

Above *Now becoming difficult to read, the inscription at the centre of the Blériot aeroplane shape gives details of the cross-Channel flight.*
Right *Charles Stewart Rolls, the first man to cross the Channel and return in a single flight (June 2 1910) as he stands on Dover seafront.*

the smashed tip was to find its way, in the fullness of time, to the safe-keeping of the RAeS in Hamilton Place, London. At first it seemed that there was no one to greet the Frenchman as he climbed stiffly down and looked about him at the open meadow which the site was at that time, 200 feet above sea level and in sight of Dover Castle. Before long he was joined by Police Constable Stanford who had seen the landing and by M Fontaine who had brought along a very large Tricolour to wave his friend in through the gap in the cliffs. It was not long before a swelling crowd gathered, made up of early morning walkers, soldiers from the Castle and a Customs Officer who, perplexed how to describe the new means of transport entered it as a 'yacht' with Blériot as the 'master and owner'.

Today one may look in vain for any sign of the position as it appears in old photographs for the area is now heavily overgrown and the outline of the monoplane, perpetuated in stone, lies in a little clearing marked by a Tricolour windsock and a finger post. A fading plaque records the details of the flight. To the north-west is another memorial, this time a statue in memory of an Englishman, none other than the Hon Charles Stewart Rolls, the partner of Royce of impeccable motor car fame. The figure here stands and appears to look out to sea as is proper for a pioneer who crossed the 22 miles twice in a single flight on June 2 1910, both his points of departure and return being Dover. The bravery of such a flight becomes evident when it is realised that his pilot's certificate was numbered '2' by the Royal Aero Club which issued it and he had qualified only on March 8 of the same year. The aircraft was a Short Wright biplane and he did not live long to enjoy the fruits of success for he was

Above *The former sea-plane shed at Dover. Aircraft were taken across the road from here to the shore for launching.*

Left *The Royal Flying Corps Memorial on Swingate Downs, Dover.*

Below *A close up of the plaque.*

The ROYAL FLYING CORPS contingent of the 1914 BRITISH EXPEDITIONARY FORCE consisting of Nos 2, 3, 4, and 5. Squadrons flew from this field to AMIENS between 13 and 15 August 1914 killed a month later at the Bournemouth Meeting and flying was suspended for the day as a mark of respect.

Behind the statue, as we look at it now, the pointed roof of a low building is visible. This is an early seaplane shed later pressed into use for other purposes although it is quite easy to see how it gave immediate access to the water in the earlier days of flying. Four years after the epic flight of Louis Blériot, crossing the Channel had become accepted and it was during a heatwave in August that the odd collection of BE2As, RE8s—and Blériot Monoplanes, that made up the RAC contingent of the British Expeditionary Force, left for France from the open space now under the shadow of the modern radar masts on Swingate Downs. The spot is marked by a granite column with a plaque recording the fact. An annual ceremony of commemoration is still held here, to mark the departures on August 13, 14 and 15 1914.

About four months later, aerial warfare came to Dover and to England in the form of the same attack, for the first bomb to ever fall on British soil dropped here on the night of Christmas Eve. It was discovered in a Mr Tyrrell's kitchen garden and did not ignite as it was not a high-explosive but a tow-wound incendiary type. It was preserved and ended up as an exhibit in the Dover Market Place Museum. This missile had been dropped by a German seaplane and it was very similar aeroplanes to this which were the equipment of the Dover seaplane base, Short 180s, and these were brought across the road from the shed already mentioned behind the Rolls statue. Once on the shore they were floated and took off on convoy supervision patrols to guard against hunting U-boats and it was not long before the numbers were swollen by Short 225s. Later, Sopwith Tabloids and FBA flying-boats which formed the very earliest equipment were withdrawn. The Station was finally closed in 1920 but not before it had enjoyed its moment of history when squadrons from here formed part of the support for naval forces taking part in the Zeebrugge raid on April 23 1918.

A few miles along the road at Deal may be found another memorial, a short distance south of Walmer. Even the road at this point has its aviation connections for it is on record that here landed a BE2C—in order that the pilot might ask the way to Swingate Downs. The child who supplied the necessary information was rewarded by an immediate flight to the destination! The Station on the main road was known as Hawkshill Down and the spot commemorated is a few yards distant. It was shared by the Royal Naval Air Service and the Royal Flying Corps and was one of the earliest aerodromes in the country. The first tenants were the Sopwith Camels of No 3 Squadron sent there for a rest after their part in the heavy fighting in the Arras area to which the squadron later returned before going to Bray Dunes in January of the next year. Meanwhile Flight Lieutenant Kerby downed two Gothas on August 12 and 22 flying from here.

In December 1917, a month after No 3, the Camels of No 4 Squadron RNAS came to this spot, a unit that had shared the Bray Dunes Aerodrome with No 3 at an earlier period. Both were to return to Walmer in March 1918, the same month that No 8 Squadron flew in at the end of a refit period. They were to return to the continent almost immediately, however, where they had to burn their aircraft when the enemy overran their base although, two days later, they were re-equipped.

The pilots involved in their stays at the Walmer air station were personally known to Viscount Beauchamp and it was the Viscountess who presented the wooden memorial. It was first erected near to the old hangars with which her husband had been familiar as Lord Warden of the Cinque Ports but, in 1952, it was moved to a more prominent position by the Walmer-Kingsdown footpath where it was re-dedicated by Air Marshal Sir Aubrey Ellwood. After a period when it was maintained in good order it fell into decay, following a period of confusion as to who was responsible for its maintenance. It was then rescued by Lieutenant Colonel Priestley-Bell, warden of the Seaside Camp for London Boys at the time, so that it still survives as a memorial to the 15 pilots whose names are inscribed on it and who died between 1914 and 1918.

Opposite Swingate Downs on the north side of the road from Dover to Deal (Swingate is on the south) lies the site of Gunston training station, purpose-built for landplane instruction. At the time this station was opened the equipment consisted of Maurice and Henri Farmans, the trainers of the day, but these were replaced by Avro 504Ks and Sopwith 1½ Strutters and later Camels. Apart from the instructional role of Gunston Road it also provided a base convenient for machines to be ferried direct to France and in common with training stations in the London area some aircraft were earmarked for interception duties, at first against airships and later Gothas. The aerodrome was finally closed down in 1920.

On the Dover to Folkestone road, and about three miles inland from the former, lies Capel. In its time this was one of the best-known airship stations from which flew the lighter-than-air craft to patrol the Straits, hunting submarines and keeping a watchful eye on convoys in the Channel. Much of the former site is now built over but there are former crewmen who remember when the sheds were erected on the foundations originally laid as part of a garage for motor coaches.

The first airships based here were all ex-army types including such famous names as the old Beta, Delta and Gamma vessels. Before long these were replaced by the larger SS-types and an immediate problem was posed. This was due to the fact that the new class was some five feet larger than Beta for which the main shed had been built with the result that the SS-type would not pass under the lintel. Clearly, something had to be done about this seemingly ridiculous problem because, had it not been for the height of the entrance, the hangar was otherwise perfectly adequate. The solution was not long in coming to Naval minds. This took the form of simply lowering the floor of the shed to allow for the increased depth of ship and car and the provision of a slope from the outside ground level. This fact, by a curious set of circumstances brought about by the layout at Capel, introduced a new word into the English language.

At first it may appear to be the mere repetition of a well-known legend that the word 'blimp' was born from the sound that seemed to be produced when a Naval commander tapped an inflated airship envelope with his finger nail. This is certainly true as far as it goes but what is not immediately realised is that in ordinary circumstances the height of the fabric above ground would have made this impossible. The way in which it came about is explained when the sloping entrance is considered for, on this particular day, *SS12* was parked on the ramp so that the envelope could easily be reached from the higher, normal ground level and the officer had only to reach out his hand as he passed to touch the silver fabric and add a new word to the dictionary!

A journey round the coast northwards would bring the visitor from the Dover

An SS-type airship.

vicinity into the very interesting area of the Thames Estuary and part of an envisaged array of concrete 'sound mirrors' along the Thames Estury. Three had been built soon after the end of the First World War and consisted of concave cement slabs with a microphone at their centre intended to pick up the sounds of incoming aircraft which echoed from the surface, and one each had been built at Hythe, Dungeness and Dover. Trials with these were conducted as early as 1928 which seemed to indicate that the 'mirrors' were not effective beyond a range of seven miles, although this was gradually extended to 15 miles from the coast under ideal weather conditions. As a result a whole system of these concrete spherical segments were planned, 25 along the greater part of the south-east coast as far as Swanage with a further 13, three of them giant ones of 2,000-foot diameter, along the Thames Estuary.

From Dover a trip along the coast in the other direction and past Dungeness would bring one to the holiday resorts of the south coast proper and only a few miles before reaching Hastings there lie the open sands of Camber. Surprisingly these number among Great Britain's pioneer aviation spots, for flying took place from here as early as 1910. The airman making these attempts was Alec Ogilvie who was the owner of one of the very first versions of the Wright 'Flyer' biplane which had been built by the Short brothers. They introduced the modification of fitting an NEC motor, although the chain drive of the original design was retained. Housed in a single hut on the wind-swept dunes, the preparation of the biplane could be a lengthy task for, without a wheeled undercarriage, it had to be hauled into position with the aid of a windlass on a dolly. Despite these limitations a flight was achieved of 140 miles non-stop from here and of five minutes under four hours duration.

A short distance away Hastings brings one aeronautically forward a few years in time for it is little realised that here was based one of the RFC Training Wings in 1917 with the flying field outside the town to the east at Fairlight above the

RAF Officer Cadets parade at St Leonards-on-Sea, Sussex, in 1918.

cliffs. This establishment was operated in two parts, one in the neo-Tudor buildings later to become a chest hospital on the high ground of St Leonards overlooking Grosvenor Crescent. The second part was in Cornwallis Terrace which backs onto the railway line, a fact which the cadets found handy in the bitter winter of 1917/18 when they were in the habit of 'borrowing' coal at dead of night from the Railway Company's yard at the rear.

By May 1918 the training establishment was well established and were even issuing their own journal called *Roosters and Fledglings* edited from the Royal Victoria Hotel, St Leonards, by an appropriately named Corporal B. Macdonald-Hastings. Even King George V came here and finally took the salute at a march-past down the road beside West Marina Gardens with the sea at his

Hastings cadets in white 'PT' shirts spell out the name of the new Service.

Roosters and Fledglings *published this drawing 'Two Strings to Her Bow'*
to mark the formation of the Royal Air Force.

Gloster Meteor T7, WL345, *on display near the seafront of St Leonards-on-Sea.*

back and some rather domestic-looking potted plants near his feet at the base of the dais. From the quaintly-named district of Bo-peep one can journey westward and still find items of aviation interest. One of these lies within a few hundred yards of the earlier spot and defies being missed—outside a motor showroom there is a complete Gloster Meteor T7, *WL345,* mounted on a concrete plinth.

The next point of interest on the south coast dates from a period some 40 years away from the days when the jet trainer probably first appeared on the St Leonard's sea-front, in fact from the high summer of 1940 just after the attack in the early evening of August 15 on the satellite fighter station at Croydon. The Messerschmitt 110 raiders were chased off across southern England and it was a staff machine that fell to the guns of a Hawker Hurricane just about three miles north-west of Bexhill-on-Sea at School Farm Hooe. The time was exactly 6.50 pm when the machine, Werk Nr *3339,* piloted by the Gruppe Technical Officer, Leutnant Koch, with Unteroffizier Kahl, came down to a wheels-up landing over the open land, delivering its crew into captivity. One marking which attracted much attention was the small reproduction of the Gruppe insignia on the nose which consisted of a red map of the British Isles within a yellow gun-sight. It was noted that southern Ireland was carefully omitted.

The next port of aeronautical call is firmly back on the coast and is only a few miles distant at Eastbourne. At once the name conjures up in the minds of aviation enthusiasts the memory of the Aviation Company. This was named after the town which produced, under sub-contract, many of the Avro 504 trainers of the 1914-18 period but it also means more for, in the early years of this century, the district also boasted an aerodrome. It sprang directly from the experiments of a young bicycle mechanic who designed and constructed his own aeroplane near Wilmington in 1911. At the same time a form of flying field came into being under the guidance of another adventurous young man and soon the two were operating in concert in the summer of the same year, the field was officially opened on December 1.

The pair responsible were Victor Yates, the aeroplane builder, and Frederick

Fowler. When the new aeroplane school was established and Victor became its instructor he was helped, one feels certain, by his having a captive pupil on the site, for Fowler was attempting to teach himself the art of the pilot. In this he was duly certificated and in a flush of enthusiasm decided to fly in the first Blériot Monoplane from Beaulieu. Unfortunately he lost his way in the fog and carried on out over the Solent into which he finally crashed but remained afloat (due to the thoughtful provision of a buoyancy bag) until rescued by the crew of a Naval launch which deposited him at Calshot, revived him with rum and provided dry clothes.

On St Anthony's Hill the flying business continued to flourish despite the fee of £30 for the course. The place swiftly became a mecca for the foremost flying men of the day so that it was only a matter of time before the Navy became interested in the form of Commander Samson. This was in the early spring of 1914 and, with the outbreak of war, seven months later, it became inevitable that the field would be requisitioned. This took place on the first day of the following September. Soon a programme of enlargement was put into effect, the civilian staff, meanwhile, were absorbed into the RNAS and, from this, Frederick graduated to the position of Flight Commander and found himself at Dunkirk.

With the creation of the new Service called the Royal Air Force, Fowler had become a Major but in common with so many others the 1920s and the years to follow found him giving joy-rides from such south coast resorts as the beach at Brighton while the Eastbourne Aerodrome slowly faded away and was finally closed down to fall a victim to the housing developer almost at once. Despite

This Messerschmitt 110D of Stab/Erpr Gr 210 was shot down by two Hurricanes of No 111 Squadron at 6.50 pm on August 15 1940, three miles north of Bexhill-on-Sea. This photograph has been marked by the wartime censor.

Europa, *on loan to the GLC, is seen conducting a survey in 1975.*

such a sad ending to an enterprise that began in the heady atmosphere of the Edwardian afterglow there was another air station which has somehow remained better remembered in the area, at least by historians. This was Polegate which lay in the area behind the present Coppice Avenue at Little Wilingdon and was opened in the summer of 1915 to act as a base for the airships which were used for Channel patrol. This choice was effected partly by the ease of access afforded by the London to Eastbourne road, not to mention the railway line with its adjacent station, and partly because a fold in the Downs gave a measure of protection from the wind.

Exactly a year after its opening, Polegate was the scene of experiments with Guardian Angel parachutes reminiscent of those destined to take place later in London from Tower Bridge but in this case not only was a drop of greater length attempted (2,000 feet) but also one from as much as a little over 1,000 feet. There was, however, one important difference in that these trials were made with weighted dummies and not live parachutists, although for the furtherance of the tests a volunteer was substituted. It is a reflection on the attitudes to safety of the day that the London trials a full year later were still grudgingly called 'experimental' despite this pioneer work over Sussex.

After four years service which took it past the day in November 1918 when the local church bells rang in joy at the announcement of the Armistice, the station was finally closed in 1919. It did not take long for the signs of service habitation to vanish and the pre-fabricated structures were soon gone, only to re-appear in other parts of the district as farm structures. However one, more solidly built, was to remain as the sole reminder that under the bricks and mortar of a housing estate lay the land that had once been one of the best-known airship stations in Dover Command. The dirigibles from here had been familiar sights all over south-eastern England, sometimes as far inland as London itself. It was almost as an echo of a past age that, in August 1975, another and more modern airship was to appear in the skies when the Goodyear tyre company loaned the *Europa* to the Greater London Council for a brief

After its emergency landing the engines of Hannibal *were removed in preparation for the machine to be dismantled near Tonbridge, Kent* (E.T. Maidment).

spell. Its duties included a survey of the Thames and Wandle Valleys and the six-hour mission provided information on a wide variety of subjects from traffic and pedestrian movements to aerial pollution.

Scattered over the area of south-east England which lies between the coastal belt and central London is a very wide choice of aeronautical connections, some known to a wider number of enthusiasts, many no more than part of the local lore. Perhaps an example of the latter is the fact that few people, even those who live in the area, seem to realise that in the tiny burial ground of Newtimber, a few miles inland from Brighton, there is the grave of one of the crew-members of the Fairey Long-Range Monoplane, a type which was to leap into the headlines in 1929 and again in 1933.

Further inland, and two years before the last long-distance attempt, the people, whose homes were close to the Tudeley-Five, Oak Green area, close to Tonbridge, had an unexpectedly close look at an airliner of the period when one came down on a very small piece of clear land at Tatingbury Farm. This was not in any sense a crash because the machine in question was *Hannibal*, one of the HP42s that enjoyed the reputation for completing ten million miles without causing hurt to a single passenger.

These majestic biplanes were a familiar sight for they passed directly over the town, making their way to and from the continent en route for Croydon. But on a hot Saturday with a slight haze over the Downs on August 8 1931 it became clear that the HP42 was going to do something considerably more spectacular than just pass over, for it was at only 800 feet, considerably less than its accustomed height. As it came lower it was plain to see the reason for this, for the propellers of only two of the four motors were turning under power since one had fractured in flight and the broken tip had flown off and put the other motor out of action. The situation left the pilot, Captain Dismore, little alternative than to put down as quickly as possible and with the relatively open land below he made the most of the opportunity.

In a graceful arc over the Paddock Wood section of the neighbouring railway line the big machine swept over Tonbridge and came in across the road where obstructions seemed almost to spring up in its path: first the roof of the local vicarage, causing a moment's disquiet to the Rev H. Capel, whose home it was, then a cottage and finally a telegraph pole. The last seemed to appear at the most critical moment and there was no hope of avoiding it so the sturdy timber was snapped off at a blow. The tail unit of the *Hannibal* went next but, despite such predicaments as these, the airliner came to a smooth halt without further damage, and the somewhat shaken, although completely unharmed, 18 passengers alighted and thanked their lucky stars for the skill and courage of such men as Dismore. It was not long before the district was the centre of attraction, first for the sightseers and newsmen, then by the engineers from Imperial Airways. The latter brought scaffolding, lorries and tools and slowly the machine was dismantled, motors first. By the following Wednesday the main part of the wings and fuselage was taken away so that the pride of the Imperial Airways fleet made an ignominious return, piecemeal and on lorries, by night to Croydon where it was repaired and served for several further years.

As one crosses from the less populated areas to the denser parts of Surrey, the connections with aviation continue to multiply. Often they are of a very small nature and of a type which must be repeated in hundreds if not thousands, not only in the south-east but throughout the British Isles. A typical example is the report in September 1970 of the discovery of Nazi buttons and coins near the spot where a Dornier crashed into White Hill, Bletchingley, in 1940. Minor finds such as these will be made for several hundreds of years still to come. Also there are those moments in time that have scarcely left a mark, even on men's minds, such as the forgotten fact that, in 1918, dismantled aircraft were stored behind the Royal Enclosure at Ascot. There can be very few, too, who now know that at much the same time a de Havilland machine got into difficulties over Twickenham and crashed through the roof of a house there, or that a BE2C snagged the telephone wires at Hampton Court and finally lodged in the trees beside the river nearby.

It would be wrong, however, to return too swiftly to Middlesex for there is still much to find in the less urbanised part of Surrey, and a case in point is to be found near Box Hill. In the little churchyard of Mickleham there is buried, in a grave surmounted by the unusual feature of a picture of a Blériot Monoplane, Graham Gilmour, one of the country's most distinguished pioneer flyers. He was killed in an accident on February 17 1912, participating in a cross-country flight from Brooklands to London, and it was he who delighted the crowds and press but infuriated officials when he gave an aerial display over the 1911 University Boat Race.

Yet although it is for escapades such as this, and the show he put on over the Henley Regatta, that Gilmour is remembered. He was no mere playboy but a serious pioneer aviator and highly accomplished pilot. Typical of his work was when he entered in the Brooklands to Brighton Air Race. He did not win this, taking over an hour and a half to complete the course in his Bristol Boxkite aircraft, but this was the sort of exhibition which won him invitations to demonstrate aeroplanes on behalf of various manufacturers. This makes his death, which occurred over Richmond Park when part of the wing structure of the machine he was flying collapsed in freak wind conditions, a double loss to aviation.

A Whitehead advertisement, showing Sopwith Pups, published in 1919.

This same area of Surrey, where it borders the Thames, was also where the original Whitehead works was established at Townsend Road, Richmond. During the period 1914-18 the firm became one of the best-known aircraft contractors, perhaps due to John Whitehead's flare for publicity which included the description of Sopwith Pups produced by the company as 'Whitehead Fighting Scouts!'. These machines were of Surrey origin since the parent firm, Sopwith Aviation Company, haled from Canbury Park Road, Kingston, where the former skating rink was taken over for aircraft manufacture.

A far less well-known aviation connection may be found elsewhere in the same town for it was at 37 Burton Road that the McCudden family lived. This included the remarkable James, holder of the VC, DSO, MC, and MM, and his no less remarkable brothers and sisters, William, Anthony, Maurice, Mary and Kate. When the last of the boys who had spent their childhood in the big old house here eventually died, at the end of 1934, No 3 Squadron RAF from Kenley marked his funeral with military honours.

The connections with aeronautics continue in a widening circle from here for only a few minutes' flying time away was, at one time, the manufacturer of amongst the best-known airscrews in the world, Lang of Weybridge. In both directions the atmosphere of aeronautica spreads out from here and among the latest is the establishment of Thorp Park at Chertsey. Here in 400 acres is a vast cross-section of the many items of heritage belonging to a maritime nation but this is interpreted in the widest sense so that, in replica form, flying and static aeroplanes are included. They embrace a wide spectrum and include machines from both sides which fought over the trenches of Flanders as well as attaining

Above *Finished in the markings once used by Ernst Udet, this Fokker D VII replica is flown by the Leisure Sport organisation.*
Below *Sopwith F1 Camel replica from Thorp Park.*
Above right *The plaque at the old Post Office at Cranbrook, Kent, which marks the first Royal Observer Corps Operations Room in 1925.*

The first
Royal Observer Corps
Operations Room
was located in this building
in 1925

THIS PLAQUE WAS ERECTED TO COMMEMORATE
THE 50th ANNIVERSARY OF THE CORPS
BY SERVING AND FORMER MEMBERS
OF THE ROYAL OBSERVER CORPS

MCMLXXVI

achievements in the racing field. In flying replica form may be seen the Supermarine S5 and the outright winner of the Schneider Trophy for this country in 1931, a full-size mock-up of the S6B having been added in 1980.

In the opposite direction there still remains the open space at Bromley, back once more in Kent, where the last of the 'lone Spitfires', leading the Battle of Britain fly-past in 1959, had to be put down hurriedly on the Oxo Company's playing field when *SL574* developed engine trouble. The pilot was Air Vice-Marshal Harold Maguire, DSO, who, faced with the problem of a fighter with a failing motor, was forced to intrude on the match then being played between the Old Hollingtonians XI and that of the Oxo Sports Club.

There is nothing to mark the spot now, unlike the site of an historic operations room, namely that of the first reporting centre of the Observer Corps. It had still to receive its prefix of 'Royal' gained for its prowess in the summer of 1940 when Major General Ashmore, CB, CMG, CVO, organised a series of trials on the reporting of aircraft movements as part of the air exercises of 1924. On a building now occupying the site of the old Cranbrook, Kent, Post Office, the present occupants, the International Stores, permitted a memorial plaque to be unveiled by Mrs B. Ashmore, at a ceremony performed after a service in St Dunstan's Church on May 9 1976.

But now our journey has come almost full cycle in that we have returned to London's Kent. A short distance away are the dormitary areas of London's Surrey. From there it is no great way to perhaps the district around Victoria where our investigation began. Even now, having looked at so much, we have done little more than scratch the surface of the vast subject that covers aviation in London and the south east.

In the mid-1930s Romford, Essex, boasted Hillman Airways which advertised a return ticket to Paris for £5.10.0.

Index

A naval airship, R26, *passing over the City of London in 1918* (Imperial War Museum).